T0280929

Novel Developments for Sustainable Hydropower

Peter Rutschmann · Eleftheria Kampa ·
Christian Wolter · Ismail Albayrak ·
Laurent David · Ulli Stoltz · Martin Schletterer
Editors

Novel Developments for Sustainable Hydropower

 Springer

Editors
Peter Rutschmann
Hydraulic and Water Resources Engineering
Technical University of Munich
Munich, Germany

Christian Wolter (iD)
Leibniz Institute of Freshwater Ecology
and Inland Fisheries
Berlin, Germany

Laurent David
Institut Pprime, Pole ecohydraulique
OFB/IMFT/Pprime Poitiers
CNRS University of Poitiers
Poitiers, France

Martin Schletterer (iD)
Department of Hydropower Engineering,
Group Ecology
TIWAG – Tiroler Wasserkraft AG
Innsbruck, Austria

Institute of Hydrobiology and Aquatic
Ecosystem Management
University of Natural Resources and Life
Sciences Vienna (BOKU)
Vienna, Austria

Eleftheria Kampa (iD)
Ecologic Institute
Berlin, Germany

Ismail Albayrak
Laboratory of Hydraulics, Hydrology
and Glaciology (VAW)
ETH Zurich
Zurich, Switzerland

Ulli Stoltz
Hydraulic Development
Voith Hydro Holding GmbH & Co. KG
Heidenheim, Germany

ISBN 978-3-030-99137-1   ISBN 978-3-030-99138-8   (eBook)
https://doi.org/10.1007/978-3-030-99138-8

This Springer imprint is published by the registered company Springer Nature Switzerland AG
The registered company address is: Gewerbestrasse 11, 6330 Cham, Switzerland

Acknowledgments

The work and contributions to this book, and the printing fully, were mainly funded by the FIThydro project, which was supported by the European Union's Horizon 2020 research and innovation programme. We also thank the Swiss National Science Foundation (SNSF), which funded the budgets of the Swiss partners. 26 partners have contributed to the success of FIThydro with their working groups and the people involved. Their scientific contributions, which have been incorporated into this book, are hereby thanked and acknowledged. A big thank you goes to the external expert advisory board of FIThydro (Martina Bussettini, Colin Bean, Robert Fenz and Chris Katopodis) who enriched the scientific work of the project with their expertise. We gratefully acknowledge the scientific support and work provided by the French Biodiversity Office (Office Français de la Biodivesité, OFB). Also, the scientific findings presented would not have been possible without the practical work in 16 Testcases provided by the hydropower operators of the consortium (Bayerische Elektrizitätswerke GmbH, BKW Energie AG, Hidroerg – Projectos Energéticos, Lda., Limmatkraftwerke AG, C. H. SALTO DE VADOCONDES, S. A., SAVASA, Statkraft AS, TIWAG – Tiroler Wasserkraft AG, Uniper Kraftwerke GmbH, VERBUND Hydro Power GmbH) and the French operators Ondula and Société Hydroélectrique de Gotein. Their generous support during the fieldwork with personnel, equipment and operational adjustments is particularly highlighted here. Finally, this book was only possible through the help and efforts of many people. Hannah Schwedhelm, Lea Berg and Hany Abo El-Wafa from the coordinators team at the Technical University of Munich deserve special mention. Without their commitment, this book would not have come about in time and in the present quality.

This project has received funding from the European Union's Horizon 2020 research and innovation programme under grant agreement No 727830.

About This Book

Hydropower is probably the oldest source of energy in the world with roots going back to the 1st to 2nd millennium B.C., when the power of water was known and mechanically used by the advanced civilisations of ancient China, Egypt and Mesopotamia. The first inventions that converted mechanical energy into electrical energy by means of reaction turbines date back to the eighteenth century. Around the beginning of the twentieth century, turbines such as Kaplan, Francis and Pelton, generated electricity with already very high efficiency. Most of the inventions in hydropower come from Europe, and even today, Europe supplies the largest share (approximately two-thirds) of hydropower equipment to the world. Electrical energy is one of the prerequisites for industrialisation, which in many countries enabled agricultural societies to develop into modern industrial nations.

For many years, society used electricity from hydropower uncritically because the economic advantages of hydropower were considered unbeatable. Hydropower plants and their components, such as turbines, were optimised for profit. The environmental adverse effects of the technology were not of great concern to the society of that time. Since hydropower plants are long-lived—50–100 years—today's society is confronted with old plants whose adverse effects on the environment are now better understood and also critically addressed. However, mitigation of the negative impacts of existing plants is much more difficult than considering appropriate mitigation strategies in the design and planning of new power plants.

Today we know much more about the effects of hydropower on ecology and especially on fish. In particular, we now know quite well—especially with the new knowledge that will be presented in this book—how negative effects can be mitigated. Nevertheless, economic efficiency is still usually a much more important aspect in planning than eco-friendliness. Therefore, the fact that this book shows example turbine hill-charts defining optimal operation with respect to minimising fish harm rather than maximising economic efficiency is an exception rather than standard practice.

In 2000, the European Commission enacted the European Water Framework Directive (WFD), which for the first time defined obligatory ecological standards for our water bodies, their fauna and flora. Because of the WFD, new requirements on the operation and design of hydropower plants were imposed, which influenced their economy.

The importance of renewable energy production to reduce CO_2 emissions was already recognised at the end of the last century, and the European Union set itself the target of generating 12% of energy consumption through renewable energy by 2010. The first Renewable Energy Directive (RED) established in 2009 a mandatory 20% share of EU energy consumption from renewable energy sources by 2020. The RED was revised in 2018 increasing the renewable energy target to 32% by 2030 and just recently, in summer 2021 a revision was suggested targeting 40% by 2030 which means a doubling of the share by the end of the decade. Even though the potential conflict between the RED and the WFD was foreseeable, it was for a long time not directly addressed. So far, the RED only set financial incentives for small hydropower, which produces only a small percentage of the total hydropower production of the EU.

According to the latest data, hydropower production in the whole of Europe amounted to 674 TWh in 2020 (Source: Eurostat the Statistical Office of the European Commission). In the 27 EU Member States, i.e. excluding some of the largest European producers such as Norway, Turkey and Switzerland, the share of hydropower in the EU_2020 still amounts to 364 TWh. The production of electricity in the 27 Member States from hydropower – excluding pumped storage production – was for the first time slightly surpassed by wind energy in 2019 (Source: Eurostat). Hydropower, however, plays an important role in the supply of electricity from renewable energy sources, not only because of the high proportion of electricity generated with hydropower, but also because of the stable safeguarding of a base load and the short time to balance seasonal storage of energy in reservoirs and pumped storage plants.

The EU WFD, which aims at protecting water bodies, set environmental targets for standing and flowing waters which among the renewable energy sources only affected hydropower. The WFD requires the assessment of water bodies and aims to achieve a so-called good ecological status for all of them. The WFD also included requirements for non-deterioration and the unimpeded passage of rivers for fish and other aquatic organisms. These points represented a major turnaround for the protection of flowing waters in particular and thus for hydropower as a whole. Passability requires functioning fishways at transverse structures and a bypass for fish around turbines to facilitate safe downstream passage around hydropower plants. In addition, the EU Biodiversity Strategy 2030 aims to put biodiversity on the road to recovery by 2030 for the benefit of the climate and the planet as a whole. With regard to hydropower, all of these aspirations contain conflicting goals and require a balancing of which of the goals and conditions to prioritise (for more details, see Chap. 1). This conflict of objectives has been recognised by the EU, but has not been resolved. However, at the national level, some countries, such as Austria, weigh the benefits of hydropower in terms of renewable energy production against environmental harm. In others, such as Germany, energy production is not seen as an explicit social benefit. In accordance with research needs recognised by the EU, the FIThydro project, which was the starting point for the results presented in this book, tried to overcome this conflict of objectives by searching for best-practice solutions that increase the protection of aquatic species while taking into account the highest possible cost-efficiency in terms of plant costs as well as generation losses at hydropower plants.

With the introduction of the WFD in 2000, the production of renewable energy from hydropower without pumped storage in the EU has settled at a constant level (source: Eurostat). In some countries, there was a slight increase between 2004 and 2020, in other countries production even decreased slightly due to the regulatory changes. While new construction of classic hydropower plants stagnated, production from pumped storage plants increased significantly. Existing plants were expanded and new plants were built to integrate the volatile renewable energies from wind and solar into the renewable energy mix.

Since the introduction of the WFD significantly influenced the annual growth of the hydropower share, the question arises as to whether it is possible to implement ecological improvements in a more cost-efficient manner for hydropower and create win-win situations, both for ecology and economy. As an example, experience shows that society is divided on the issue of hydropower acceptance (Chap. 2) and that opinions are so entrenched that compromises are often seemingly impossible. In water rights procedures, opponents of hydropower often tend to maximise and operators to minimise environmental flow. In most cases, the final result is a regulation that sets the residual flow to a certain volume of water usually with a distinction between a summer half-year and a winter half-year flows. However, it is easy to understand that such a supposed "best" compromise solution is rarely optimal and often open to challenge.

This book is an outcome of the FIThydro project and summarises the novel and important results and findings originating from this EU H2020 project. The FIThydro project itself responded to an EU call (LCE-07-2016) that focused on the impacts of hydropower on river ecology and the possibilities to mitigate such negative impacts. The call addressed the fact that hydropower has great importance for the European renewable energy targets, but that solutions are still required to meet the environmental targets of the EU, to preserve clean energy and to mitigate the negative effects on the environment, especially through innovation and new approaches.

The keywords of the call for proposals were river ecology, self-sustainable fish populations, habitat improvement, identification of fish species most at risk and suitable methods, models and devices, and reliable, quantitative figures on fish mortality in turbines. To achieve these goals, the call requested that high-quality datasets should be re-analysed and innovative conclusions drawn. All these aspects needed to be elaborated and studied at existing hydropower plants at several locations in the EU in order to draw practical and transferable conclusions for operators, planners and decision-makers.

The FIThydro consortium consisted of 26 partners covering all the necessary areas in an interdisciplinary and even transdisciplinary way. It included practically equal parts of science, industry, operators and planners. Fish biologists worked together with ecologists and engineers; sociologists and economists were also part of the team. This type of cooperation was innovative for most of the partners, and also had an intensity and closeness that was unique for such a project. We started as a team that brought a lot of experiences and knowledge with us and thought that we were able to contribute to solving many problems. Despite the achievements, we all realised that aspects from different disciplines are highly complex, interlinked and that much is not yet understood and further research is needed. Ultimately, this reality is very much at odds with current planning, where simplified knowledge and buzzwords often dominate decisions and even more so the understanding and attitudes of different stakeholders.

The FIThydro project identified eight main objectives to be achieved. Namely

1. Bringing together all disciplines related to hydropower
2. Assessing the response and resilience of fish populations in HPP affected rivers
3. Environmental impact assessment and species protection
4. Improving fish and fisheries impact mitigation strategies using conventional and innovative cost-efficient measures
5. Enhancing methods, models and tools to cope with EU obligations
6. Identifying bottlenecks of HPPs and deriving cost-efficient mitigation strategies
7. Risk-based Decision Support System (DSS) for planning, commissioning and operating of HPPs
8. Enhancing problem awareness and objectiveness of policy implementer, NGOs and the public.

In addition to the plans for "Communication and Dissemination" and "Project Management", these overarching goals were worked on in five work packages:

The first work package dealt with "Fish population development in HP affected environments", and the second with "The appraisal of existing solutions, models, tools and devices to assess (the) self-sustained fish population(s) at the Testcase HPP in each of the four regions". The third work package addressed "The innovation of solutions, models, tools and devices to assess self-sustained fish population(s) at the Testcase HPP in each of the four regions" and the fourth addressed "Cost-effective management strategies to improve the development of self-sustained fish populations at existing and new HPPs". The outcomes of these work packages were integrated in the fifth work package that was concerned with "Stakeholder involvement & decision-support systems".

16 Testcases, on which the research and innovative methods were undertaken, formed the focal point of the project. All project innovations were not only literature-based or research-based, but they also had real-world testing and a high practical relevance via the Testcases. This required that the partners from the different disciplines to work together closely and to deepen communication and knowledge exchange. The 16 real Testcases were brought into the project by the operator partners at the proposal writing phase. At each of these Testcases, there were one or more real challenges that had to be solved to mitigate the ecological impact of the respective HPP. Challenges were related to upstream

and downstream fish migration, flow alterations, habitat loss and sediment transport issues. Each of these overall challenges was subdivided into several sub-challenges.

As these Testcases were brought into the project relatively randomly and could not cover all areas of the real challenges, scenario modelling was also carried out at Testcases, in which virtual "sandpit games" were undertaken to solve the question: What if...? For example, what if a different turbine was installed, or what if a safe downstream bypass for fish was in place? In this way, it was possible to artificially create situations that were necessary for generalised statements at all the given Testcase sites.

In the course of the FIThydro project, 40 deliverables of which 21 were technical deliverables and further 7 content documents were created, submitted and are accessible on the Internet. In particular, a FIThydro wiki was also created as a living document that will make all project content accessible for years to come (www.fithydro.wiki). Much of this knowledge was gathered from literature reviews, i.e. it is content that was not created in the course of the project and is therefore not unique or new. This book, however, focuses on the new content that has emerged through FIThydro. It does this in a way that allows an interested layperson, rather than a scientist, to gain a quick overview of key results. The book can be a beginner's literature for an expert who wants to gain quick access to a new topic.

Due to the genesis of the content presented, this book does not have a classical chapter structure resulting from an overarching logic in which the reader is guided from one contribution to the next without gaps in the text and content. Rather, the individual contributions reflect the logic and also the restrictions of the FIThydro project. Although they are assigned to thematic blocks, they do not claim to have worked through these blocks comprehensively and without gaps. The individual contributions of the various working groups contain the following topics:

Chapter 1 introduces the policy framework on EU and national level relevant for the planning and operation of hydropower and of measures to mitigate environmental impacts. Chapter 2 evaluates the costs of hydropower mitigation measures or more precisely their cost ranges. Chapter 3 provides an overview of public perception of hydropower projects and describes methods for studying public acceptance. Chapter 4 briefly outlines site- and constellation-specific direct and indirect impacts of a hydropower scheme primarily on fishes. Chapter 5 deals with the conventional upstream fish passage technologies developed for safe and effective fish migration at run-of-river hydropower plants. Chapter 6 presents an agent-based model which has been developed and implemented in the fish habitat model CASiMiR identifying fish migration corridors and applying behavioural rules. Chapter 7 deals with solutions for downstream fish migration using physical barriers and fish guidance structures with narrow bar spacing and bypass systems. In contrast, Chap. 8 presents results using wide bar spacing and the effect of fish behavioural barriers for guiding fish downstream and also presents the innovative Curved Bar Racks, which additionally minimise head losses. Chapter 9 focuses on turbine passage as one option for downstream migration of fish at hydropower plants and quantifies the impact of turbine passage. Chapter 10 presents ways of influencing fish survival rates during turbine passage and shows the positive effects of these. Chapter 11 deals with fish and their damage rates in Archimedes screws and presents results of a study conducted at a very large Archimedes screw hydropower station. Chapter 12 adapts the hydropeaking COSH-TOOL developed and applied in Scandinavia for salmon and for native Iberian cyprinids in Portugal. Chapter 13 investigates the effect of hydropeaking at the upper Inn River with the CASiMiR habitat modelling tool which was extended to account for temporal changes and the speed of these changes and their effects on juvenile habitat conditions. Chapter 14 presents restoration work at the Inn River in Bavaria, supporting sustainable fish populations by improving and extending remaining habitats and reproducing historically available habitat elements in hydropower affected rivers. Chapter 15 introduces a series of tools and guidance to assess environmental hazards of hydropower in particular on fish, to enhance assessing cumulative effects from several hydropower schemes and to enable informed decisions on planning, development and mitigation of new and refurbished hydropower schemes. Chapter 16 draws conclusions and gives an outlook to future research fields.

Munich, Germany   Peter Rutschmann
                                                              peter.rutschmann@tum.de

Reference

Eurostat, the Statistical Office of the European Union, Energy statistics—quantities, annual data (nrg_quanta), SHARES tool 2020 (https://ec.europa.eu/eurostat/de/web/energy/data/shares), 30.11.2021.

Contents

Contributors

Ana Adeva-Bustos SINTEF Energy Research, Trondheim, Norway

Ismail Albayrak Laboratory of Hydraulics, Hydrology and Glaciology, ETH Zurich, Zurich, Switzerland

Raf Baeyens Research Institute for Nature and Forest (INBO), Brussels, Belgium

Isabel Boavida CERIS – Civil Engineering Research and Innovation for Sustainability, Instituto Superior Técnico / University of Lisbon, Lisbon, Portugal

Robert M. Boes Laboratory of Hydraulics, Hydrology and Glaciology, ETH Zurich, Zurich, Switzerland

Francisco Javier Bravo-Córdoba ITAGRA – University of Valladolid, Palencia, Spain

David Buysse Research Institute for Nature and Forest (INBO), Brussels, Belgium

Damien Calluaud Institut Pprime, Pole ecohydraulique OFB/IMFT/Pprime, CNRS University of Poitiers, Poitiers, France

Omar Carazo-Cea ITAGRA – University of Valladolid, Palencia, Spain

Julie Charmasson SINTEF Energy Research, Trondheim, Norway

Ludovic Chatellier Institut Pprime, Pole ecohydraulique OFB/IMFT/Pprime, CNRS University of Poitiers, Poitiers, France

Johan Coeck Research Institute for Nature and Forest (INBO), Brussels, Belgium

Dominique Courret OFB/IMFT/Pprime, Office Français de La Biodiversité, Toulouse, France

Ian G. Cowx Department of Biological and Marine Sciences, Hull International Fisheries Institute (HIFI), University of Hull, Hull, UK

Laurent David Institut Pprime, Pole Ecohydraulique OFB/IMF/Pprime, CNRS University of Poitiers, Poitiers, France

Carlos Escudero-Ortega ITAGRA – University of Valladolid, Palencia, Spain

Juan Francisco Fuentes-Pérez ITAGRA – University of Valladolid, Palencia, Spain

Ana García-Vega ITAGRA – University of Valladolid, Palencia, Spain

Franz Geiger Ecohydraulic Consulting Geiger, Wallgau, Germany

Juergen Geist Aquatic Systems Biology, Technical University of Munich, Freising, Germany

Holger Gerdes Ecologic Institute, Berlin, Germany

Francisco Godinho Hydroerg – Projectos Energéticos Lda, Lisbon, Portugal

Atle Harby SINTEF Energy Research, Trondheim, Norway

Mandy Hinzmann Ecologic Institute, Berlin, Germany

Tobias Hägele sje – Ecohydraulic Engineering GmbH, Stuttgart (Vaihingen), Germany

Eleftheria Kampa Ecologic Institute, Berlin, Germany

Myron King Department of Biological and Marine Sciences, Hull International Fisheries Institute (HIFI), University of Hull, Hull, UK

Ianina Kopecki sje – Ecohydraulic Engineering GmbH, Stuttgart (Vaihingen), Germany

Georg Loy VERBUND Innkraftwerke GmbH, Töging am Inn, Germany

Richard A. A. Noble Department of Biological and Marine Sciences, Hull International Fisheries Institute (HIFI), University of Hull, Hull, UK

Joachim Pander Aquatic Systems Biology, Technical University of Munich, Freising, Germany

Ine S. Pauwels Research Institute for Nature and Forest (INBO), Brussels, Belgium

Armin Peter FishConsulting GmbH, Olten, Switzerland

Gérard Pineau Institut Pprime, CNRS University of Poitiers, Pole ecohydraulique OFB/IMF/Pprime Poitiers, Poitiers, France

António Pinheiro CERIS – Civil Engineering Research and Innovation for Sustainability, Instituto Superior Técnico / University of Lisbon, Lisbon, Portugal

Johannes Radinger Leibniz Institute of Freshwater Ecology and Inland Fisheries, Berlin, Germany

Walter Reckendorfer VERBUND Hydro Power GmbH, Vienna, Austria

Peter Rutschmann Hydraulic and Water Resources Engineering, Technical University of Munich, Munich, Germany

Francisco Javier Sanz-Ronda ITAGRA – University of Valladolid, Palencia, Spain

Martin Schletterer Department of Hydropower Engineering, Group Ecology, TIWAG – Tiroler Wasserkraft AG, Innsbruck, Austria;
Institute of Hydrobiology and Aquatic Ecosystem Management, University of Natural Resources and Life Sciences, Vienna, Austria

Matthias Schneider sje – Ecohydraulic Engineering GmbH, Stuttgart (Vaihingen), Germany

Nils Schoelzel FishConsulting GmbH, Olten, Switzerland

Nicole Smialek Aquatic Systems Biology, Technical University of Munich, Freising, Germany

Ulli Stoltz Hydraulic Development, Voith Hydro Holding GmbH & Co. KG, Heidenheim, Germany

Jeffrey Tuhtan Department of Computer Systems, School of Information Technologies, Tallinn University of Technology, Tallinn, Estland

Jorge Valbuena-Castro ITAGRA – University of Valladolid, Palencia, Spain

Ruben van Treeck Leibniz Institute of Freshwater Ecology and Inland Fisheries, Berlin, Germany;
Ecology of Fish and Waterbodies, Institute of Inland Fisheries Potsdam-Sacrow, Potsdam, Germany

Michael van Zyll de Jong Department of Biological Sciences, University of New Brunswick, Saint John, NB, Canada

Terese E. Venus Agricultural Production and Resource Economics, Technical University of Munich, Fresing, Germany

Lisa Wilmsmeier FishConsulting GmbH, Olten, Switzerland

Christian Wolter Leibniz Institute of Freshwater Ecology and Inland Fisheries, Berlin, Germany

Policy Framework for Hydropower Mitigation

1

Eleftheria Kampa ⓘ

1.1   Introduction

Policies are crucial in determining and improving the state of our environment. Policies set goals (including targets, indicators and time frames) while policy instruments are the specific means or measures to translate the policy intent into action (Jacob et al. 2019). In discussions on the sustainability of hydropower production, knowledge of the currently existing regulations at different levels is indispensable (Bunge et al. 2003) in understanding the framework conditions for decisions on impact-mitigating measures. Hydropower plants play an important role in the production of renewable energy and in the reduction of CO_2 emissions. At the same time, hydropower can have a range of negative effects on the flow of rivers, the habitats of fish and aquatic organisms, as well as on fauna and flora species that depend on river and lake ecosystems for their survival (EC 2018). Therefore, hydropower is at the cross-road of goals and instruments of different policy fields on energy/climate, water and nature. All these different policies need to be taken into account in a balanced way considering synergies and trade-offs when planning and implementing mitigation actions for hydropower.

In the European Union (EU), several EU and national policies set ecological and environmental requirements on hydropower plants. These include in particular policies for the protection of nature and water resources. Further, the planning and operation of European hydropower plants takes place in the framework of policies that promote and support the

E. Kampa (✉)
Ecologic Institute, Berlin, Germany
e-mail: eleftheria.kampa@ecologic.eu

© The Author(s) 2022
P. Rutschmann et al. (eds.), *Novel Developments for Sustainable Hydropower*,
https://doi.org/10.1007/978-3-030-99138-8_1

production of renewable energy. Illustrated with examples, this chapter introduces the policy and legislative framing for hydropower in the EU and in selected European countries as well as its implications for the planning of mitigation measures and the realisation of more sustainable hydropower.

1.2   Policy Framework for Hydropower Mitigation

This section reviews key EU and national policies with requirements for mitigating the ecological impacts of hydropower, illustrated with examples from a policy survey of eight European countries (Kampa et al. 2017). In addition to EU policy objectives, national policies specify and operationalise the policy framework in which hydropower operators need to mitigate the impacts of hydropower production. Policy requirements need to be considered in the decision-making processes of hydropower operators and authorities, both in the relicensing process of existing hydropower plants (HPPs) and the licensing of new HPPs.

The main policies that are relevant to the planning and operation of hydropower plants address renewable energy and climate change, water resource protection, water resource infrastructure, nature and biodiversity protection, fisheries, invasive alien species and project impact assessment (Fig. 1.1).

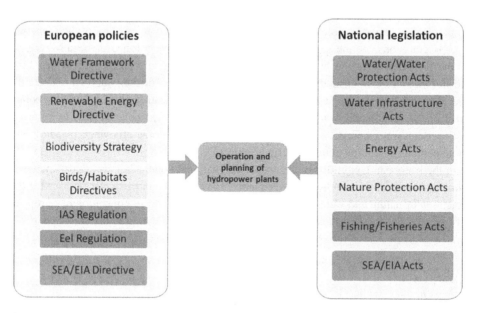

Fig. 1.1 Broad categories of European and national policies relevant to the operation and planning of hydropower plants

Due to the focus on EU policies and national policies in European countries, global processes are not addressed in this section. It is noted though that a number of international agreements are relevant to sustainable hydropower production, such as the 2030 Agenda for Sustainable Development, the Convention on Migratory Species (CMS), the Convention on Biological Diversity (CBD) and the United Nations Framework Convention on Climate Change (UNFCCC).

1.2.1 European Policies

The planning, operation and mitigation of impacts of hydropower plants need to be aligned with the key objectives of EU policy on water, energy as well as biodiversity protection. The recently adopted European Green Deal (EC 2019) sets a framework for aligning the objectives of these different policies, in view of the Green Deal aims for a "climate neutral" Europe and for enhanced protection of European ecosystems and biodiversity.

As hydropower production takes place in the aquatic environment, the objectives of the EU Water Framework Directive (WFD) need to be a primary consideration. Since its adoption in 2000, the WFD has been a strong driver for restoring aquatic ecosystems and river continuity. The WFD is the key policy taken into account in European countries when modifying the licensing procedures for new hydropower plants and when revising licenses of existing plants. The Directive's main aim (Article 4(1)) is to prevent deterioration of status and achieve good status of all EU waters, including surface and groundwater, by 2015 (at the latest by 2027). For surface waters, the Directive distinguishes between good ecological and good chemical status. Ecological status is "good" when the values for biological quality elements of surface waters (fish, benthic invertebrates, aquatic flora, phytoplankton) deviate only slightly from undisturbed conditions. Hydromorphological and physico-chemical parameters are supporting elements to the biological quality elements for classifying the status of water bodies.

A river basin management plan has to be established for each river basin district defined under the WFD in a cyclical process every six years, which involves the development of a programme of measures to tackle significant pressures on water bodies, including those from hydropower activities in the form of barriers on rivers, hydrological alterations and abstractions.

So far, progress in achieving the WFD aim of good status has been slow, with only around 40% of surface waters being in good ecological status or potential in the second river basin management plans of 2015 (EEA 2018). Significant pressures from hydropower production affect approximately 9000 surface water bodies (6% of total surface water bodies) (EEA 2021). In addition, hydropower is the most common reason for designating heavily modified water bodies according to Article 4(3) of the WFD applicable to approximately 6000 water bodies (half of these water bodies being in Norway)

(EEA 2021). Water bodies can be designated as heavily modified and have lower environmental objectives, when it is not viable to remove physical modifications such as a dam and particular water uses and public interests stand in the way of extensive restoration of the water bodies in question.

Concerning the development of new hydropower plants or modifications to existing plants in Europe, Article 4(7) of the WFD has to be taken into account. Article 4(7) makes new infrastructure projects possible, only if certain strict conditions are met, i.e. there are no significantly better environmental options, the benefits of the new infrastructure outweigh the benefits of achieving the WFD environmental objectives and all practicable mitigation measures are taken to address the adverse impact on the status of the water body.

Next to water policy, EU policy on biodiversity and nature protection is another major layer in the policy framework of hydropower production in Europe. At locations affected by hydropower, Natura 2000 site provisions for the protection of certain species and habitats need to be taken into account. Furthermore, any plan or project that could affect a Natura 2000 site should be subject to an assessment procedure to study these effects in detail, based on Article 6(3) of the Habitats Directive. This is relevant for new hydropower projects and for upgrades or modernizations of existing hydropower plants. Except for the provisions within the Natura 2000 network, the Habitats Directive and the Birds Directive concern the protection of certain species across their entire natural range within the EU, i.e. also outside Natura 2000 sites. These provisions need to be taken into account by operators of hydropower plants, especially on rivers harbouring migratory fish species that are listed in Annex IV of the Habitats Directive, such as the European sea sturgeon (EC 2018).

Progress in achieving EU nature protection objectives has to this day been slow, as only 15% of habitat assessments identified good conservation status between 2013 and 2018. Key pressures on European habitats and species include modifications to hydrological flow, physical alterations including barriers and hydropower installations (EEA 2020).

The new EU Biodiversity Strategy adopted in 2020 includes targets for restoring freshwater ecosystems that are also relevant to activities of hydropower production. The Biodiversity Strategy targets include the restoration of at least 25,000 km of rivers into free-flowing rivers by 2030 and the restoration of degraded ecosystems, like floodplains and wetlands. In addition, water abstraction and impoundment permits should be reviewed to implement ecological flows in order to achieve good status or potential of all surface waters and good status of all groundwater by 2027 in line with the WFD (EC 2020a).

Equally important to the EU water and nature protection policy framework is the EU policy agenda for renewable energy and climate change mitigation and adaptation. Energy production from hydropower plays a key role for the EU to meet its energy needs and climate mitigation targets in the future. The revised EU Renewable Energy Directive (EU 2018) established a new binding target for the EU on the share of energy from renewables of at least 32% of the Union's gross final consumption in 2030, with a clause for a possible upwards revision by 2023. EU countries are required to draft 10-year National Energy & Climate Plans for 2021–2030, outlining how they will meet the new 2030 targets for renewable energy and energy efficiency.

There are significant differences between EU countries in terms of the extent to which hydropower is used in their renewable energy mix. This is highly influenced by geographic conditions, climate, precipitation patterns, the availability of affordable energy supply alternatives, as well as institutional capacities and technical competences (Kampa et al. 2017). To meet the EU renewable energy targets for 2030, some European countries will increase the use of hydropower for energy storage as well as energy production. For example, there are plans to expand hydropower pumped storage in Austria, the Baltic States and Portugal (IHA 2019).

Other EU policies which are relevant to the planning, operation and mitigation of impacts of hydropower plans are the EU Eel Regulation, the EU Regulation on Invasive Alien Species as well as the Strategic Environmental Assessment and Environmental Impact Assessment Directives (SEA/EIA).

The EU Eel Regulation (1100/2007) requires the establishment of measures for the recovery of the stock of the European eel (which is a species impacted by the presence of hydropower), the identification and definition of eel river basins and the set-up of Eel Management Plans to reduce anthropogenic mortalities and improve the escapement of the silver eel to the sea. Progress between 2007 and 2020 though on the achievements of the Eel Regulation has been limited. Overall biomass and escapement levels of silver eel have not yet significantly improved. Implementation of the Eel Regulation needs considerable improvement, especially when addressing non-fisheries related anthropogenic mortality, such as impacts from hydropower and dams (EC 2020b).

Also the EU regulation on Invasive Alien Species (in force since 2015) is relevant to hydropower plants, as the development of hydropower installations can create new connections between river systems, leading to the spread and dispersal of various aquatic organisms. In implementing this EU Regulation, Member States may include requirements in their hydropower authorisation procedures related to preventing the spread of alien species.

Finally, the Environmental Impact and Strategic Environmental Assessment Directives (EIA and SEA) are of particular relevance to the consenting of new projects in the hydropower sector. An environmental impact assessment is mandatory for dams and other installations that hold back or permanently store water, where a new or additional amount of water held back or stored exceeds 10 million cubic meters (Annex I projects of EIA Directive). Most installations for hydropower production in Europe though are Annex II projects under the EIA Directive, i.e. projects for which Member State authorities must first determine, in a procedure called "screening", if projects are likely to have significant effects, taking into account certain criteria, before deciding on whether a project will be made subject to the EIA procedure. The SEA Directive is more relevant for planning hydropower at a larger and more strategic scale rather than at the level of individual hydropower projects. A Strategic Environmental Assessment may be relevant for national plans and programmes on the development of hydropower including a more appropriate siting of future developments to avoid potential areas of conflict such as in Natura 2000 sites.

1.2.2 National Policies

National legislation defines the regulatory setting for planning and operating hydropower plants and for planning mitigation measures in a particular country. According to a recent review of hydropower-related national legislation by Kampa et al. (2017), the following are the main types of national acts that are used to regulate hydropower plants (HPP): water/water protection acts; water infrastructure acts; energy acts; nature protection acts; fishing/fisheries acts and environmental impact assessment acts.

National regulatory settings differ from country to country. The review by Kampa et al. (2017) carried out in the context of the FIThydro project gives a detailed account of the relevant acts in place in eight European countries: Norway, Sweden, France, Portugal, Spain, Germany, Switzerland and Austria.

Some of the relevant acts date back to the early 1900s or even earlier, for instance the Watercourse Regulation Act in Norway (1917), the Law on fish and fisheries in France (1865) and the Rivers Fishing Act in Spain (1942). However, most of the acts on water resource management, biodiversity protection and renewable energy are more recent and have been formulated under the influence of key EU policies. Certain recent amendments to national legislation have high relevance for the operation or the commissioning of hydropower plants. For example, the Austrian National Water Act revised in 2011 (WRG Novelle 2011) made several mitigation measures such as upstream continuity measures and ecological minimum flow an obligatory requirement. Also in Switzerland, new developments and environmental requirements for hydropower plants are driven by the amendment of the Waters Protection Act in 2011 (Federal Act on the Protection of Waters 2011). The amended Swiss Water Protection Act requires the mitigation of impacts from hydropeaking, the remediation of impairments of the bedload regime and fish migration restoration until 2030 as well as implementation of measures for river revitalisations and improvement of the morphology until 2090 (Schweizer 2017). In Sweden, a number of revisions to the Environmental Code that entered into force in 2019 required the development of a national plan for the revision of hydropower plant licenses with greater focus on environmental goals.

National policy largely defines how environmental requirements for mitigation measures are set for hydropower plants. This mainly takes place in the context of the authorisation procedures for HPP (new authorisation procedures and revision or renewal of authorisations already in place). When planning mitigation measures for new or existing hydropower plants, the mitigation requirements based on legislation, or other types of recommendations (e.g. best-practice guidelines or technical standards) or specific decisions by permitting authorities need to be reviewed and taken into account.

Mitigation requirements for hydropower plants can be distinguished for the following broad domains of environmental improvements at hydropower plants: upstream fish migration, downstream fish migration, flow conditions, hydropeaking, sediment transport, and habitat enhancement.

Requirements for hydropower plants to mitigate the impacts of disrupted upstream fish migration and modified flow conditions are usually based on legislation. In the review of Kampa et al. 2017, relevant requirements in legislative form were found in the majority of the eight countries of the study (Table 1.1). To mention but a few examples: In Austria, ensuring ecological continuity is compulsory except outside of the natural fish zone and very near to natural existing barriers. In Germany, the federal states have set specific technical and hydraulic requirements for upstream fish migration measures. In France, the maintenance of minimum flow is an obligation since 2006 with the requirement to implement minimum flow values by 2014. Concerning minimum flow requirements, overall different methods are used by different countries for its determination (Ramos et al. 2017).

In a number of countries, there is still a lack of requirements based on law to mitigate impacts related to sediment transport, hydropeaking impacts and downstream fish migration, mainly due to knowledge gaps and a lack of proven measures that need clarification through research or pilot studies. For these types of impacts, countries often follow a case-by-case approach when defining mitigation requirements (see overview of situation in eight European countries in Table 1.1). There are exceptions though, for instance, in Germany, several federal states set specific requirements (e.g. on protection screens) for fish protection and downstream migration in ordinances and guidelines (Kampa et al. 2017).

Overall, the EU WFD and recent revisions of national policies are strong drivers for modifying the authorisation procedures for new hydropower plants as well as for revising authorisations of existing plants. The possibility to carry out revisions to permits can help ensure that hydropower plants become environmentally sounder and that state-of-the-art mitigation measures are implemented. However, a large number of hydropower plants in European countries were built prior to modern environmental laws with inadequate environmental requirements in place and no mechanisms to revise them.

In several European countries, a transformation of the regulatory framework is taking place to overcome such barriers. Next to specific technical requirements for impact mitigation, the duration of hydropower plant permits plays a key role in the options available for revisions and inclusion of new mitigation measure requirements. Due to recent changes in environmental legislation and social pressure, permit duration for hydropower plants has in general been reduced in many countries, although the duration of permits still generally differs between new and existing plants.

Table 1.1 Overview of the presence of mitigation requirements for hydropower in eight countries reviewed in the FIThydro project (based on the review by Kampa et al. 2017)

Environmental improvement domains	In legislative form	Recommendations (guidance approaches)	Case-by-case definition	No requirements
Upstream fish migration	Austria, France*, Germany, Norway, Portugal, Spain		France, Portugal	Sweden
Downstream fish migration	France*, Germany, Switzerland, Spain	Germany	Austria (new HPPs), France, Norway, Portugal	Austria (existing HPPs), Sweden
Flow conditions	Austria, France, Germany, Norway, Portugal, Switzerland, Spain	Portugal	Germany, Sweden	
Hydropeaking	Austria (new HPPs), Germany, Switzerland, Spain		France, Germany, Norway, Sweden, Portugal	Austria (existing HPPs), Sweden
Sediment transport	Austria (new HPPs), France, Switzerland, Spain		France, Germany, Portugal	Austria (existing HPPs), Sweden, Norway
Habitat enhancement	Austria (new HPPs), France (new HPPs), Germany, Switzerland, Spain	Norway	Austria, France (existing HPPs), Portugal	

*In France, mitigation requirements for upstream and downstream fish migration and sediment transport are based on legislation, if the river is listed in "list 2" that is a list of rivers where it is necessary to ensure the movement of migratory fish and sediment transport. If the river is not in "list 2", mitigation requirements are defined case-by-case

In Germany, new hydropower plants are usually granted permits up to 30 years, while older plants have ancient rights (often indefinite concessions), which are permits that were granted to operators or installations when the Water Act first came into force in 1960. The permit conditions under ancient rights are often environmentally inadequate from today's perspective and it is difficult for authorities to compel these operators to modernise. A permit revision is normally only needed if the turbine power is planned to be increased. However, water authorities have recently been getting stricter and asking for mitigation measures. This is also in the case of indefinite concessions, especially when the rivers in question are priority water courses for fish conservation (e.g. Programmgewässer Lachs) (Kampa et al. 2017).

In Sweden, according to a new national plan, all existing hydropower licenses will be reviewed over the next 20 years. Unlimited concessions will no longer be granted, with a maximum for new concessions of 40 years. This will involve placing greater emphasis on mitigation of impacts, including setting minimum environmental flows and the installation of fishpasses (Swedish Agency for Marine and Water Management 2019). Also in Norway, the environmental terms of licensing conditions are reviewed after 50 years (and after 30 years for larger hydropower plants built after 1992). The environmental terms of licenses in Norway that are revised in this context typically include setting minimum flow, requirements for physical habitat improvements, increased continuity, and enhancement of qualities that can be important for the water users (recreation, fishing, etc.). Overall, although the licenses are usually unlimited for publicly owned entities, the environmental terms are revised at regular intervals (Kampa et al. 2017).

In France, although the authorisation procedures have not directly been adapted to the WFD, the definition of mitigation measures is now more ambitious to preserve or restore the good ecological status of streams and rivers. In Spain, an adaptation of existing authorisations depends on the specific permit regime, the water plan of the district and jurisprudence related to determined cases. Overall though changing existing permits is complicated and is bound to produce legal challenges where existing rights of concession holders are affected. Also in Portugal, authorisations for existing hydropower are not yet required to be adapted to WFD requirements (Kampa et al. 2017). Therefore, the level of progress made is diverse across Europe and key steps still need to be taken for an ambitious approach for more environmentally sound hydropower across the continent.

In general, mitigation requirements for new and for existing HPP do not differ substantially, if there is an option to revise existing permits. In case the permit of an operating HPP runs out, in all eight countries reviewed by Kampa et al. (2017), the same conditions as for new authorisations apply in the process of permit renewal. This means that mitigation measures may be required for existing HPP, even where none were required before.

Regulations relevant to HPP authorisation may outline aspects that should be considered in addition to environmental conditions, when setting mitigation requirements. Consideration of cost (dis-)proportionality, cost balancing and limits on the economic feasibility of the HPP are the most commonly additional aspects taken into account in authorisation procedures. Therefore, the implementation of mitigation measures is usually not decided solely upon ecological criteria.

1.3   Conclusion

Policy requirements need to be considered carefully when taking decisions on mitigation measures to reach policy objectives, both in the relicensing process of existing HPPs and the licensing of new HPPs. The overarching framework is set by key EU legislation in particular the Water Framework Directive, EU policies on nature and biodiversity protection and the EU agenda for renewable energy and climate change mitigation and adaptation. The EU policy framework is further specified and operationalised by national legislation that provides the regulatory setting for planning and operating hydropower plants and mitigation measures in each European country. The review of policy in eight European countries in the FIThydro project has shown how recent changes in environmental legislation as well as social pressure have generally reduced permit duration for hydropower plants and how the WFD and other national policy revisions have been modifying authorisation procedures for hydropower at the benefit of implementing mitigation measures. The impacts of hydropower for which mitigation is most commonly required by legislation include the disruption of upstream fish migration and the modification of flow conditions. For other types of impacts, in particular related to sediment transport, hydropeaking and downstream fish migration, relevant legislative requirements for mitigation are largely missing due to uncertainties and the need for more research on the effectiveness of mitigation measures.

References

Bunge T, Dirbach D, Dreher B et al (2003) Hydroelectric power plants as a source of renewable energy—legal and ecological aspects. Federal Environmental Agency (Umweltbundesamt), Berlin

EC (European Commission) (2018) Guidance on the requirements for hydropower in relation to EU Nature legislation. Publications Office of the European Union, Luxembourg

EC (European Commission) (2019) Communication from the Commission to the European Parliament, the Council, the European Economic and Social Committee and the Committee of the Regions—The European Green Deal, COM(2019) 640 final

EC (European Commission) (2020a) Communication from the Commission to the European Parliament, the Council, the Economic and Social Committee and the Committee of the Regions—EU Biodiversity Strategy for 2030. Bringing nature back into our lives. COM(2020a)380 final

EC (European Commission) (2020b) Commission Staff Working Document. Evaluation of Council Regulation (EC) No 1100/2007 of 18 September 2007 establishing measures for the recovery of the stock of European eel. SWD (2020b) 36 final

EEA (European Environmental Agency) (2018) European waters: assessment of status and pressures 2018, No 7/2018. European Environment Agency, Copenhagen

EEA (European Environmental Agency) (2020) State of nature in the EU: results from reporting under the nature directives 2013–2018, No 10/2020. European Environment Agency, Copenhagen

EEA (European Environmental Agency) (2021) Drivers of and pressures arising from selected key water management challenges—a European overview, No 9/2021. European Environment Agency, Copenhagen

EU (2018) Directive 2018/2001 of the European Parliament and of the Council of 11 December 2018 on the promotion of the use of energy from renewable sources. OJ L 328:82–209

Federal Act on the Protection of Waters (2011) Waters Protection Act (Gewässerschutzgesetz, GSchG) of 24 January 1991, Amended in 2011. https://www.fedlex.admin.ch/eli/cc/1992/1860_1860_1860/de. Accessed 3 May 2021

IHA (International Hydropower Association) (2019) Region Profile: Europe. https://www.hydropower.org/region-profiles/europe. Accessed 19 Dec 2017

Jacob K, Mangalagiu D, King P, Rodríguez-Labajos B (2019) Approach to assessment of policy effectiveness. In: Ekins P et al (eds) Global environment outlook 6: healthy planet, healthy people. Cambridge University Press, Cambridge, pp 273–282

Kampa E, Tarpey J, Rouillard DJ, Stein DU, Bakken TH, Godinho FN, Leitão AE, Portela MM, Courret D, Sanz-Ronda FJ, Boes R and Odelberg A (2017) D5.1—Review of policy requirements and financing instruments (for fishfriendly hydropower). FIThydro Project Report. https://www.fithydro.eu/deliverables-tech/

Ramos V, Formigo N, Maia R (2017) Ecological flows and the water framework directive implementation: an effective coevolution? Eur Water 60:423–432

Schweizer S (2017) Mitigation of hydropeaking in the hasliaare—selection of measure(s)—technical aspects—monitoring. In: Presentation at the CIS Workshop on GEP inter-comparison case studies on water storage, Federal Ministry for Sustainability and Tourism, Vienna, 13–14 February 2017

Swedish Agency for Marine and Water Management (2019) Towards sustainable hydropower in Sweden. https://www.havochvatten.se/en/eu-and-international/towards-sustainable-hydropower-in-sweden.html. Accessed 25 Apr 2021

WRG Novelle (2011). Änderung des Wasserrechtsgesetzes 1959. Bundesgesetzblatt Nr. BGBl. I Nr. 14/2011. https://www.ris.bka.gv.at/eli/bgbl/I/2011/14. Accessed 3 May 2021

Costs of Ecological Mitigation at Hydropower Plants

2

Terese E. Venus, Nicole Smialek, Ana Adeva-Bustos, Joachim Pander, and Juergen Geist◉

2.1 Introduction

The costs of fish-related mitigation measures can play an important role in determining which measures are adopted, yet there is relatively little publicly available information about this aspect. While the majority of the literature focuses on environmental impacts and mitigation strategies, there have only been a few studies about costs. For example, Nieminen et al. (2017) reviewed general economic and policy considerations for mitigation measures facilitating fish migration. They outlined several suggestions for simultaneously improving sustainable hydropower production and supporting migratory fish, including shifting the emphasis from technology to environmental standards and considering multiple values of migratory fish (e.g. consumption, recreation, tourism, aquatic

T. E. Venus (✉)
Agricultural Production and Resource Economics, Technical University of Munich, Freising, Germany
e-mail: terese.venus@tum.de

N. Smialek · J. Pander · J. Geist
Aquatic Systems Biology, Technical University of Munich, Freising, Germany
e-mail: nicole.smialek@tum.de

J. Pander
e-mail: joachim.pander@tum.de

J. Geist
e-mail: geist@tum.de

A. Adeva-Bustos
SINTEF Energy Research, Trondheim, Norway
e-mail: ana.adeva.bustos@sintef.no

© The Author(s) 2022
P. Rutschmann et al. (eds.), *Novel Developments for Sustainable Hydropower*,
https://doi.org/10.1007/978-3-030-99138-8_2

food webs and ecosystem functioning). Further, Venus et al. (2020a) estimated cost trade-offs between fish passage migration and hydropower in over 300 European case studies. They found that nature-like fish passages tend to incur fewer overall costs and power losses than technical designs. Finally, Oladosu et al. (2021) compiled costs of mitigating environmental impacts in the United States and showed that environmental costs vary significantly by type of hydropower project and mitigation measure. They also found that smaller plants tend to spend a higher relative share of total project costs on environmental mitigation. While these studies have focused on the costs of individual measures in specific case studies, they do not provide a robust overview of the magnitude of costs across different types of mitigation measures. This chapter presents an overview of the range of costs of different mitigation measures to compare available costs and their magnitudes. Further, as many mitigation measures are adopted in combinations, this chapter presents costs from two FIThydro case studies to understand cost considerations under different mitigation combinations. These case studies demonstrate how costs might be compared when multiple mitigation measures are adopted.

2.2   Cost Ranges of Mitigation Measures

As costs differ based on site-specific characteristics, it can be difficult to compare the costs from different hydropower plants. To provide an overview for policymakers of the magnitude of costs associated with different measures, this section summarizes costs from different sources and presents an overview of ranges of costs based on the following types of mitigation measures. Costs were collected directly from hydropower operators and energy producers (Vattenfall, France Hydro Electricité), researchers via a questionnaire, peer-reviewed literature and reports published by state authorities. To cover a wide range of regions, data from different regions (Europe, North America, Australia) were included. All costs were converted to Euros using the average 2010–2019 exchange rate (0.82 for USD/EUR and 1.46 for AUD/EUR) and rounded to defined increments to give a general impression of the cost dimensions rather than the specific costs of case studies.[1] The results are presented in Table 2.1.

2.2.1   Costs of Environmental Flow Measures

Environmental flow (henceforth e-flow) measures incur costs related to the flow release itself and structures used to release flow. The cost of release depends on several factors, specifically where, when and how much flow is released. E-flows can be released to the

---

[1] Minimum costs were rounded down and maximum costs were rounded up to the following increments: 1, 5, 10, 20, 50, 100, 150, 200, 500, 1000, 2000, 5000, 10,000, 150,000, 100,000, 1,000,000. If only one value is provided, the cost estimate is based on a single case study.

Table 2.1 Cost ranges for sediment management measures

		Measure	Costs (Euros)		Unit	Source
			Minimum	Maximum		
Sediment	Routing	Drawdown reservoir flushing	1	50	Per cubic meter	Rovira and Ibàñez (2007), Espa et al. (2013)
		Sediment sluicing	NA			
	Removal	By-passing sediments	NA			
		Off-channel reservoir storage	NA			
		Mechanical removal of fine sediments (dredging)	5	10	Per cubic meter	Rovira and Ibàñez (2007)
		Minimising sediment arrival to reservoir	150,000		Per Vortex tube	Personal communication (Doessegger 2020)
	Restoration in rivers	Removal of bank protection	NA			
		Removal of debris	NA			

bypassed river reaches or through the turbine. If water is not released through the turbine, it can result in power losses. E-flows are also typically not released constantly throughout the year. Instead, the specific environmental targets and regulations dictate when and how much water should be released (World Meteorological Organization 2019) or more information about how dynamic instream flows can be used to ensure the functionality of river dynamics, see Auerswald and Geist (2018) and Casas-Mulet et al. (2017). For information about other habitat forming processes as well as biological requirements for life history needs, see Acreman and Ferguson (2010), Forseth and Harby (2014) and Pander et al. (2018).

The costs associated with power losses depend on the amount and timing of water released. However, water losses ($m^3 s^{-1}$) cannot be directly converted into monetary losses. Water losses must first be converted to power losses (kWh). Then, power losses can be converted into monetary values using electricity prices. However, these prices can vary significantly based on the region, the time of year/day, inflow-conditions and the type of power market (e.g. balancing, day-ahead, reserve markets, etc.) (Pérez-Díaz and Wilhelmi 2010; Pereira et al. 2019; Ak et al. 2019). For this reason, there was limited information on the costs of e-flow measures. Especially at peak flows, it is also possible to use water for e-flow after the turbines have reached their utilization capacity (Pander and Geist 2013; Stammel et al. 2012). In such cases, the water used for e-flow does not decrease turbine productivity nor incur costs.

The cost of structures (e.g. gates) for flow release depends on the following factors: (i) retrofitting or new structure, (ii) use of the structure, (iii) location relative to the plant, and (iv) material/labour costs. If an existing structure is retrofitted for flow release, it will likely cost more than building a new structure. Further, the structure may be exclusively used for flow release or also used to preventing hydropeaking. If the structure is used for multiple purposes, it may also incur higher costs overall. Structures used to mitigate hydropeaking such as an attenuation reservoir can also be used. Costs increase relative to the size of the dam in the attenuation reservoir (Charmasson and Zinke 2011). The location of the structure relative to the plant is also important. Usually, such structures are built at the outlet of the plant (e.g., retention reservoirs, tunnels or bypasses). Finally, local conditions such as the cost of materials and labour will also affect the magnitude of costs. Once the structure is built, there may be some recurring costs in the form of maintenance (Venus et al. 2020b).

2.2.2 Costs of Sediment Management Measures

To understand the drivers of costs of sediment management measures, it is important to note that there are three main mechanisms for managing sediment: (i) flow release, (ii) temporary creation and maintenance of habitat (e.g., dredging), and (iii) permanent structures that facilitate sediment transport (e.g., vortex tube). Following the categories in Table 2.1 sediment routing mainly relies on flow release while removal and restoration in rivers require both temporary and permanent measures.

For the costs of flow release, refer to Sect. 2.2.1. When using flow for sediment management, there are a few specific considerations. Similar to other e-flow measures, costs are usually recurring and dependent on the lost volume of water. Although sediment management is primarily done to prevent damage to the turbines, it is also possible that damages occur and incur costs. Further, the timing of e-flow is important, as e-flow and dredging could be competing events.

Due to dynamic river processes, sediment can settle close to the hydropower station. Thus, the mechanical removal/placement of sediment is a temporary action, representing a recurring cost. The magnitude of the costs depends on several factors including structural requirements (i.e., size of the river, size of the facility, amount of gravel), site accessibility as well as machinery rental and labour costs. Mechanical removal (dredging) of fine sediments cost approximately 5€–10€ per m^3 in a Spanish case study (Rovira and Ibàñez 2007). In addition, sediment erosion downstream of hydropower dams can result in break-through events and also result in substantial cost. For both reasons, ensuring sediment transport through the dam is typically the target.

The costs of structures (e.g. sediment bypasses such as pressurised pipelines, tunnels, canals) tend to be non-recurring and depend on the site topography, obstacle size and shape and hydraulics of the river (Healy et al. 1989). A Vortex tube used to minimize sediment arrival to the reservoir was estimated to cost approximately 150,000€ per tube (Personal communication A. Doessegger 2020). Some recurring costs may be incurred in the form of maintenance.

2.2.3 Costs of Fish Migration Measures

Fish migration measures include both upstream and downstream measures and incur costs related to the cost of the structure itself, power loss and ongoing maintenance. In general, fish migration measures are constructed either when the hydropower plant is built (new) or added when new licenses are needed (retrofitted). When newly built with the power plant, the costs are generally much lower as all the engineering elements required are already available (Table 2.2).

The costs for restoring upstream fish migration are dependent on the size of the fish-pass (height of obstacle, length of fishpass, discharge of the fishpass), design (technical vs. nature like construction design), and material (concrete, rip-rap structures, cost of required land, etc.). Barrier removal restores the natural river flow and does not incur recurring costs. As the costs are per project, per unit costs can be calculated. Between types of fishpasses, there is a wider range of costs for pool-type and baffle passes compared to nature-like passes. This may be linked to site-specific issues. If the site is difficult to access, construction of passes with concrete may incur relatively higher costs. Nature-like passes may incur comparatively lower costs as they use natural materials (e.g. stones, vegetation, etc.) rather than concrete. However, pool-type and baffle passes may require less space and can often be designed according to standard formulas. Depending on the location, the costs of acquiring additional land may prohibit the construction of natural passes. Fish lifts, screws and locks tend to incur higher costs per project as these technologies are more complex and only preferred at hydropower plants with limited space or very high heads.

As nature-like passes may necessitate more space to overcome a higher obstacle (i.e., land acquisition costs) and cannot be standardised like technical passes (i.e. planning and construction costs), they are often thought to incur greater costs. However, in a review of European fish passage facilities, nature-like measures were found to cost less than technical measures even when controlling for the height of the obstacle and length of the pass. As nature-like fishpasses can also serve habitat functions including spawning

Table 2.2 Cost ranges for fish migration measures

		Measure	Costs (Euros)		Unit	Source
			Minimum	Maximum		
Fish Migration	Downstream	Operational measures (turbine operations, spillway passage)	NA			
		Sensory, behavioural barriers (electricity, light, sound, air–water curtains)	800	4000	Per m^3/s	Turnpenny et al. (1998)
		Fishfriendly turbines	500,000		Per turbine	Dewitte et al. (2020)
		Skimming walls (fixed or floating)	3,000		Per m^3/s	Venus et al. (2020c)
		Bypass combined with other solutions	10,000	25,000	Per m^3/s	Ebel et al. (2018)
		Fish guidance structures with narrow bar spacing	2,000	40,000	Per m^3/s	Venus et al. (2020b)
		Fish guidance structures with wide bar spacing	2,000	40,000	Per m^3/s	Venus et al. (2020b)
		Bottom-type intakes (Coanda screen)	17,000		Per m^3/s	Turnpenny et al. (1998)

(continued)

Table 2.2 (continued)

		Measure	Costs (Euros)		Unit	Source
			Minimum	Maximum		
	Upstream	Complete or partial migration barrier removal	2,000	1,000,000	Per project	California Department of Fish and Game (CDFG) (2004)
		Nature-like fishways	5,000	20,000	Per vertical meter	Rutherfurd et al. (2000)
		Pool-type fishways	10,000	100,000	Per vertical meter	California Department of Fish and Game (CDFG) (2004), Porcher and Larinier (2002), Venus et al. (2020b)
		Baffle fishways	5,000	100,000	Per vertical meter	California Department of Fish and Game (CDFG) (2004), Venus et al. (2020b)
		Fishways for eels and lampreys	600		Per meter length	Pulg et al. (2020)
		Fish lifts, screws, locks, and others	10,000	500,000	Per project	Venus et al. (2020b)
		Trap and truck	NA			

or feeding habitats, investing in nature-like solutions may be the preferable conservation action (Pander et al. 2013). For an analysis of how different factors affect costs related to fish migration measures, see Venus et al. (2020b).

Downstream migration measures tend to be less technically advanced (Porcher and Larinier 2002). As many downstream migration measures are adaptations of existing facilities at hydropower plants (screens/racks) or operational changes, there is less information about their costs. Downstream migration can be facilitated through either passive (flow release) or active (screens, sensory/behavioural barriers, other guidance structures) measures. No information on the costs of operational measures (i.e., turbine operation, spillway passage) was found in the review. This may be because they are site- and operation-specific. Sensory and behavioural barriers ranged in costs from 800 to 4,000€ per m³/s (Turnpenny et al. 1998). An example of a fishfriendly turbine (Very-Low-Head) costs 500,000€ per turbine (Dewitte et al. 2020). Skimming walls cost approximately 3,000€ per m³/s (Venus et al. 2020b). Bypasses combined with other solutions range from 10,000€ to 25,000€ per m³/s (Ebel et al. 2018). Fish guidance structures either with narrow or wide bar spacing ranged from 2,000€ to 40,000€ per m³/s (Venus et al. 2020b). A Coanda screen cost approximately 17,000€ per project (Turnpenny et al. 1998).

2.2.4   Costs of Habitat Measures

There are a variety of measures, which can be used to improve aquatic habitats in hydropower affected environments. They range from small-scale measures that address single life stages of species to the holistic restoration of ecosystem functioning (Table 2.3). In general, the more complex the restoration target, the higher the costs of mitigation (Pander and Geist 2013). Habitat mitigation measures incur costs related to (i) temporary adjustments of physical habitat and (ii) permanent construction measures. Adjustments to the flow conditions through the release of water can also improve ecosystem functioning. The magnitude of costs depends on the several site-specific factors: ecological targets, desired habitat type, degree of habitat connectivity, size of the area to be restored, materials and site accessibility (Pander and Geist 2018).

The temporary creation of physical habitat entails instream habitat adjustments such as the placement of spawning gravel, stones and deadwood as well as the cleaning of substrate. The costs of such measures are usually recurring. This is because many habitat improvements are not self-sustaining as obstacles (e.g. hydropower plants) in the river have altered natural river dynamics. Hence, these measures have to be repeated or improved over time. For example, the introduction of gravel for spawning grounds is usually needed on a yearly basis in catchments with high erosion rates (Pander et al. 2015). The restoration of habitat (e.g., construction of off-channel habitats) and shoreline habitat (e.g. restoration of the riparian zone vegetation) tends to be non-recurring.

Table 2.3 Cost ranges for habitat measures

		Measure	Costs (Euros)		Unit	Source
			Minimum	Maximum		
Habitat	Instream habitat adjustments	Placement of spawning gravel in the river	10	100	per cubic meter	Personal communication Loy (2020), Personal communication Zehender (2020)
		Placement of stones in the river	50	150	per cubic meter	Cramer (2012)
		Cleaning of substrate—ripping, ploughing and flushing	1	50	per square meter	Cramer (2012)
		Placement of dead wood and debris	10	150	per meter	Cederholm et al. (1997)
	Restoring habitat	Construction of a 'river-in-the-river'	50	5,000	per meter	Saldi-Caromile et al. (2004)
		Construction of off-channel habitats	1	100	per square meter	Evergreen Funding Consultants (2003)
	Shore-line habitat	Environmental design of embankments and erosion protection	10	150	per meter	Cramer (2012)
		Restoration of the riparian zone vegetation	1	50	per square meter	(Evergreen Funding Consultants 2003)

The costs of habitat measures are more accessible compared to other hydropower mitigation measures as they are often applied in non-hydropower contexts. However, it is important to note that in hydropower-affected environments, functional reliability of the energy system must be guaranteed and this can in turn cause higher costs for habitat measures. For example, drifting deadwood in a hydropower-affected environment is likely to be more expensive since it needs additional structures such as anchor bodies to secure it on site for safety reasons (Pander and Geist 2010, 2016).

2.3   Cost Comparisons from FIThydro Testcases

The FIThydro project studied several Testcases with different environmental targets to assess their cost-effectiveness. In this section, the costs of two Testcases are presented: Las Rives in France and Guma in Spain. While different mitigation strategies may incur costs related to energy losses and construction costs, they may also enable increased energy production.

The Las Rives hydropower plant is situated on the River Ariège in southern France in a reach home to cyprinids and salmonids. The river ecosystem is affected by hydropower as well as agricultural runoff (e.g. nutrients, pesticides). There are mitigation targets related to downstream and upstream migration as well as e-flows. Although French authorities require a specific amount of e-flow, the operator released less by agreeing with the authorities to improve downstream fish migration conditions at the plant. Specifically, the trash rack in front of the hydropower was re-designed and a new DIVE turbine was installed to increase e-flow and power production. Additionally, the plant has an alternate vertical slot pass that was integrated with a DIVE turbine to increase the attraction flow for upstream migrating fishes. As a result of mitigation, the operator increased power production and decreased fish mortality.

The Guma hydropower plant is situated on the River Duero in north-western Spain, which is home to cyprinids including some endemic ones of high conservation importance (e.g., Iberian barbell, northern straight-mouth nase, Northern Iberian chub and Pyrenean gudgeon). Dams and hydropower as well as agricultural use (e.g. irrigation) affect the ecosystem. At the plant, the operator addressed challenges related to upstream migration, spawning habitat and e-flow. For upstream mitigation, the operator installed a pool and weir fishway with a submerged notch, bottom orifice and attraction flow. Although Spanish authorities do not require e-flow, the operator ensured sufficient flow for functionality of the fishpass. Within the FIThydro project, researchers used scenario modelling to compare changes in the attraction flow at the fishway and morphological alterations between the power station tailrace and the fishpass branch. The simulated results showed that the morphological alterations and the increase of attraction flow could potentially improve upstream migration and facilitate access to the spawning areas upstream of the hydropower plant.

2.3.1   Calculating Costs of Operational Changes

Costs included operational changes (e.g. shutting down the turbines), morphological modifications (e.g. digging terrain to increase the depth) and structural solutions (e.g. trash racks). For the operational changes, annual and daily power production was calculated using the hydraulic head and turbine efficiency. This was combined with the power price to calculate the costs of increasing the e-flow and reducing the water passing through the turbines for energy production. In another case, the Short-term Hydro Optimization Program (SHOP)2 was used to calculate the loss of energy and costs of shutting down the turbines during the migration period, and from increasing the e-flow.

Energy losses were calculated by comparing the monetary values of energy production with the actual situation and production at the different hydropower plants. In both cases, the morphological and construction costs were annualized with an amortization period of 14 years and a discount rate of 5%. In Las Rives, the construction costs were in most cases higher than the power losses, considering also that the new turbine increases the production and the e-flow included in the attraction flow reduces the losses. In Guma, the morphological costs were lower, but all measures included a loss of income. However, it is important to consider that construction and morphological costs will be recovered after 14 years, but not the energy production losses.

2.3.2 Cost Comparison of Fishfriendly Measures

In Las Rives, costs of several actions related to downstream passage mitigation were compared (Fig. 2.1). These included variations of installing a new bar rack, shutting down the turbine and adding a new turbine (Fig. 2.1). Mitigation measures costs included the construction of new devices as well as income gain and losses, which consists of increasing the e-flow that is or is not used for energy production and shutting down the turbines.

(Note: Costs are ordered from lowest total cost to highest total cost. Negative costs (−) show that the measure created additional benefits, which reduced total costs of the measure.)

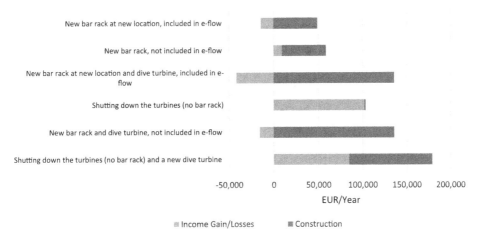

Fig. 2.1 Total costs of downstream mitigation measures at Las Rives (2.7 MW)

In Guma, costs of several actions related to different levels of e-flow and morphological changes (Fig. 2.2). Mitigation measures costs included morphological changes such as the addition of blocks from different sizes, morphological alternation of a river bed channel (by widening and shaping) and the income losses such as the increase of the e-flow from 1 to 3 respectively.

These examples from the FIThydro Testcases demonstrate how the losses associated with operational changes can be incorporated into cost comparisons for potential mitigation strategies. To improve future cost assessments of mitigation measures, it is important to make cost data publicly available as much as possible. In turn, this will improve transparency of mitigation and aid decision makers in supporting effective ecological mitigation at hydropower plants.

2.4 Conclusion

The costs of fish-related mitigation measures play an important role in their adoption. There is a wide range of costs depending on the type of measure adopted and site-specific factors. As evident from the empirical data and the experiences from the case studies, there are trade-offs between power production and mitigation, particularly when combinations of measures are adopted. However, it is also important to remember that these costs should be weighed against their ecological benefits. Specifically, they can contribute to achieving

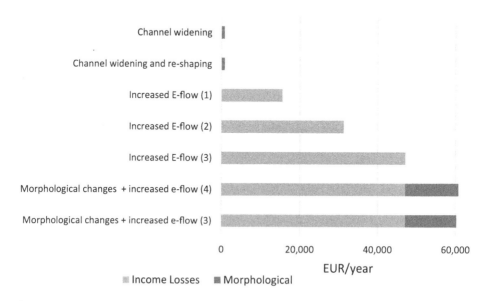

Fig. 2.2 Total costs of e-flow and morphological changes at Guma (2.25 MW)

"good ecological potential" and "good ecological status" in water bodies, a key target formulated in the European Water Framework Directive.

In light of ecological targets, managers should also consider that mitigation measures are often not self-sustaining. In such cases, managers might consider adaptive river management, which is an iterative process that responds to the dynamic river environment and improves management decisions as information is attained (Geist and Hawkins 2016). From a cost perspective, this means that costs are recurring rather than non-recurring. Similarly, monitoring is also an important part of adaptive river management. Further, environmental monitoring for hydropower has been found to be positively valued by the public and should be included in cost-benefit analyses (Venus and Sauer 2022). Thus, it is important that planners not only consider costs of the measures but also ongoing monitoring. While some critics cite monitoring costs as a disadvantage of adaptive management, investments in well-designed monitoring programs may be cost-effective compared to the costs of designing entirely new mitigation programs.

Acknowledgements We would like to thank Javier Sanz-Ronda, Ana García Vega (GEA-Valladolid University) and SAVASA. We also thank Angela Odelberg, Pål Høberg (Statkraft), António Pinheiro and Ana Quaresma (CERIS—IST, Lisboa University) for sharing information and data and for their continuous support, as well as the Confederación Hidrográfica del Duero (Duero Water Authority) and Fishing Service of the regional government of Castilla y León for their legal and technical support.

References

Acreman MC, Ferguson AJD (2010) Environmental flows and the European water framework directive. Freshw Biol 55:32–48. https://doi.org/10.1111/j.1365-2427.2009.02181.x

Ak M, Kentel E, Savasaneril S (2019) Quantifying the revenue gain of operating a cascade hydropower plant system as a pumped-storage hydropower system. Renew Energy 139:739–752. https://doi.org/10.1016/j.renene.2019.02.118

Auerswald K, Geist J (2018) Extent and causes of siltation in a headwater stream bed: catchment soil erosion is less important than internal stream processes. L Degrad Dev 29:737–748. https://doi.org/10.1002/ldr.2779

California Department of Fish and Game (CDFG) (2004) Recovery strategy for California coho salmon. Report to the California Fish and Game Commission

Casas-Mulet R, Alfredsen KT, McCluskey AH, Stewardson MJ (2017) Key hydraulic drivers and patterns of fine sediment accumulation in gravel streambeds: a conceptual framework illustrated with a case study from the Kiewa River, Australia. Geomorphology 299:152–164. https://doi.org/10.1016/j.geomorph.2017.08.032

Cederholm CJ, Bilby RE, Bisson PA, et al (1997) Response of Juvenile Coho Salmon and steelhead to placement of large woody debris in a coastal washington stream. North Am J Fish Manage

Charmasson J, Zinke P (2011) Mitigation measures against hydropeaking effects. SINTEF Rep TR A 7192

Cramer ML (2012) Stream habitat restoration guidelines. Olympia, Washington

Dewitte M, Courret D, Laurent D, Adeva-Bustos A (2020) Comparison of solutions to restore a safe downstream migration of fish at a low-head run-of-river power-plant. In: Fish Passage 2020—International Conference on River Connectivity

Ebel G, Kehl M, Gluch A (2018) Fortschritte beim Fischschutz und Fischabstieg: Inbetriebnahme der Pilot-Wasserkraftanlagen Freyburg und Öblitz. WasserWirtschaft 9/2018

Espa P, Castelli E, Crosa G, Gentili G (2013) Environmental effects of storage preservation practices: controlled flushing of fine sediment from a small hydropower reservoir. Environ Manage 52:261–276. https://doi.org/10.1007/s00267-013-0090-0

Evergreen Funding Consultants (2003) A primer on habitat project costs. Prepared for the Puget Sound Shared Strategy

Forseth T, Harby A (2014) Handbook for environmental design in regulated salmon rivers

Geist J, Hawkins SJ (2016) Habitat recovery and restoration in aquatic ecosystems: current progress and future challenges. Aquat Conserv Mar Freshw Ecosyst 26:942–962. https://doi.org/10.1002/aqc.2702

Healy KM, Cox AM, Hanes DM, Chambers LG (1989) State of the practice of sediment management in reservoirs: minimizing sedimentation and removing deposits. J Chem Inf Model. https://doi.org/10.1017/CBO9781107415324.004

Nieminen E, Hyytiäinen K, Lindroos M (2017) Economic and policy considerations regarding hydropower and migratory fish. Fish Fish 18:54–78. https://doi.org/10.1111/faf.12167

Oladosu GA, Werble J, Tingen W et al (2021) Costs of mitigating the environmental impacts of hydropower projects in the United States. Renew Sustain Energy Rev 135. https://doi.org/10.1016/j.rser.2020.110121

Pander J, Geist J (2010) Seasonal and spatial bank habitat use by fish in highly altered rivers—a comparison of four different restoration measures. Ecol Freshw Fish 19:127–138. https://doi.org/10.1111/j.1600-0633.2009.00397.x

Pander J, Geist J (2013) Ecological indicators for stream restoration success. Ecol Indic 30:106–118. https://doi.org/10.1016/j.ecolind.2013.01.039

Pander J, Geist J (2016) Can fish habitat restoration for rheophilic species in highly modified rivers be sustainable in the long run? Ecol Eng 88:28–38. https://doi.org/10.1016/j.ecoleng.2015.12.006

Pander J, Geist J (2018) The contribution of different restored habitats to fish diversity and population development in a highly modified river: a case study from the River Günz. Water 10:1202. https://doi.org/10.3390/w10091202

Pander J, Mueller M, Geist J (2018) Habitat diversity and connectivity govern the conservation value of restored aquatic floodplain habitats. Biol Conserv 217:1–10. https://doi.org/10.1016/j.biocon.2017.10.024

Pander J, Mueller M, Geist J (2013) Ecological functions of fish bypass channels in streams: migration corridor and habitat for rheophilic species. River Res Appl 29:441–450. https://doi.org/10.1002/rra.1612

Pereira JP, Pesquita V, Rodrigues PMM, Rua A (2019) Market integration and the persistence of electricity prices. Empir Econ 57:1495–1514. https://doi.org/10.1007/s00181-018-1520-x

Pérez-Díaz JI, Wilhelmi JR (2010) Assessment of the economic impact of environmental constraints on short-term hydropower plant operation. Energy Policy 38:7960–7970. https://doi.org/10.1016/j.enpol.2010.09.020

Porcher JP, Larinier M (2002) Designing fishways, supervision of construction, costs, hydraulic model studies. Bull Français La Pêche La Piscic 156–165. https://doi.org/10.1051/kmae/2002100

Pulg U, Stranzl S, Espedal E, et al (2020) Effektivitet og kost-nytte forhold av fysiske miljøtiltak i vassdrag. Bergen

Rovira A, Ibàñez C (2007) Sediment management options for the lower Ebro River and its delta. J Soils Sediments 7:285–295. https://doi.org/10.1065/jss2007.08.244

Rutherfurd ID, Jerie K, Marsh N (2000) A rehabilitation manual for Australian Streams, vol 2

Saldi-Caromile K, Bates K, Skidmore P et al (2004) Stream habitat restoration guidelines: final draft. Olympia, Washington

Stammel B, Cyffka B, Geist J et al (2012) Floodplain restoration on the Upper Danube (Germany) by re-establishing water and sediment dynamics: a scientific monitoring as part of the implementation. River Syst 20:55–70. https://doi.org/10.1127/1868-5749/2011/020-0033

Turnpenny AWH, Struthers G, Hanson P (1998) A UK guide to intake fish-screening regulations, policy and best practice with particular reference to hydroelectric power schemes

Venus TE, Hinzmann M, Bakken TH et al (2020a) The public's perception of run-of-the-river hydropower across Europe. Energy Policy 140. https://doi.org/10.1016/j.enpol.2020.111422

Venus TE, Sauer J (2022) Certainty pays off: the public's value of environmental monitoring. Ecol Econ 191. https://doi.org/10.1016/j.ecolecon.2021.107220

Venus TE, Smialek N, Pander J et al (2020b) Evaluating cost trade-offs between hydropower and fish passage mitigation. Sustainability 12:8520. https://doi.org/10.3390/su12208520

Venus TE, Smialek N, Pander J et al (2020c) D 4.3—general cost figures for relevant solutions, methods, tools and devices. FIThydro Project Report. https://www.fithydro.eu/deliverables-tech/

World Meteorological Organization (2019) Guidance on environmental flows integrating e-flow science with fluvial geomorphology to maintain ecosystem services

Public Acceptance of Hydropower

Terese E. Venus, Mandy Hinzmann, and Holger Gerdes

3.1 Introduction

In the context of hydropower development and modernisation, including the adoption of mitigation measures, public perceptions may play an important role. Critical issues in the planning stage may cause local resistance to a project and delay its completion. Hence, hydropower operators, planners and policy-makers should understand how the study of local public perceptions about hydropower may improve the planning of new projects, modernisation of existing ones and the implementation of mitigation measures as well as that criticism may be reduced by stimulating participation in the planning process. As there are a variety of methods for studying public acceptance, this chapter reviews public acceptance factors from previous hydropower studies, presents the Q-methodology and demonstrates how it can be a means for studying public acceptance and exploring subjective views on hydropower among local residents.

Using methods from the social sciences, this section on public acceptance of hydropower illustrates how public perceptions may affect the planning of hydropower plants and how hydropower operators, planners and policy-makers can improve their

T. E. Venus (✉)
Agricultural Production and Resource Economics, Technical University of Munich, Freising, Germany
e-mail: terese.venus@tum.de

M. Hinzmann · H. Gerdes
Ecologic Institute, Berlin, Germany
e-mail: mandy.hinzmann@ecologic.eu

H. Gerdes
e-mail: holger.gerdes@ecologic.eu

© The Author(s) 2022
P. Rutschmann et al. (eds.), *Novel Developments for Sustainable Hydropower*,
https://doi.org/10.1007/978-3-030-99138-8_3

understanding of these perceptions to the benefit of more socially acceptable hydropower. Examples are given from the application of the Q-methodology in a study of four European towns in hydropower-intensive regions, which revealed that different perspectives on hydropower exist among the respective local populations. For example, one perspective is that hydropower as a climate-friendly energy source is a crucial component for an energy transition. Another perspective is that hydropower potentially harms the river ecosystem. Hydropower managers should be aware of concerns and can assess public views using the Q-methodology when planning new or modernizing hydropower plants and planning mitigation measures.

3.2   Factors for Public Acceptance of Hydropower

Given growing support for the renewable energy transition, researchers have studied the public acceptance of hydropower technologies (Tabi and Wüstenhagen 2017; Venus et al. 2020). Such studies may inform decision-makers about public perceptions and help them facilitate the improved planning and implementation of policies, address resistance to renewable projects and stimulate public participation during key planning stages (Botelho et al. 2016; Ribeiro et al. 2014; Volken et al. 2019; Wüstenhagen et al. 2007).

A review of literature associated with hydropower found that there are several factors which are relevant to the acceptance of hydropower including (i) economic costs and benefits, (ii) quality of life, (iii) ecological effects, (iv) public participation and (v) energy policy and (vi) energy preferences (Venus et al. 2020). Table 3.1 provides an overview of relevant factors.

The category, "economic costs and benefits", encompasses the perceived benefits to the development of hydropower including job creation, tax revenue, stable and low-cost electricity in remote areas and energy security (Tabi and Wüstenhagen 2017; Saha and Idsø 2016). On the other hand, this also includes perceived costs such as increased energy prices and negative effects on other industries (Malesios and Arabatzis 2010; Gullberg et al. 2014).

As hydropower can affect the local population's quality of life, it is important to consider factors such as recreational opportunities, disruptions to natural scenery and habitats as well as threats to the cultural heritage of the region (Bakken et al. 2012; Botelho et al. 2016; Klinglmair et al. 2015; Saha and Idsø 2016). Acceptance can be affected by concerns about drinking water quality (Saha and Idsø 2016), noise from hydropower plants (Botelho et al. 2016) and accidents (Öhman et al. 2016). However, in a Swiss study, citizens believed the risk of accidents related to hydropower to be low (Volken et al. 2019). As some regions have a long history of hydropower, the technology has been viewed as key for the "national building process" (Lindström and Ruud 2017). As the development of hydropower has catalyzed industrialisation, the technology can be a source of pride (Lindström and Ruud 2017).

Table 3.1 Public acceptance factors for hydropower

Category	Sub-category
Economic costs and benefits	Economic development (e.g., job creation; effects on other industries such as tourism, agriculture, forestry)
	Energy prices
	Energy security
Quality of life	Recreational opportunities (e.g., fishing, bathing, boating, going for a walk)
	Flood protection
	Quality of drinking water
	Accidents
	Place attachment
	Cultural identity
	Ideal of nature
	Landscape aesthetics (i.e., natural scenery)
Ecological effects	Biodiversity and habitats
	Fish safety
	Greenhouse gas emissions
Public participation	Planning
	Profit-sharing
Energy policy	Ownership
	Subsidies
Energy preferences	Comparison to other renewables
	Modernisation
	Flexibility
	Energy storage

Views about the negative ecological effects (e.g. habitats, fish abundance and migration, etc.) of hydropower are also crucial for public acceptance (Malesios and Arabatzis 2010; Ribeiro et al. 2014; Tabi and Wüstenhagen 2017). On the other hand, hydropower can also contribute positively to climate change mitigation as it is viewed as a clean way of producing electricity (Gullberg et al. 2014; Karlstrøm and Ryghaug 2014; Klinglmair et al. 2015; Mattmann et al. 2016). It is important to note that acceptance decreases when people perceive negative ecological impacts to be greater than the benefits derived from greenhouse gas reductions (Mattmann et al. 2016). Further, environmental monitoring can also affect public views on hydropower mitigation (Venus and Sauer 2022).

The extent of public participation in the decision-making and planning process can also increase public acceptance of hydropower (Díaz et al. 2017). When stakeholders

were not able to participate in discussions about new hydropower developments in Norway, for example, acceptance decreased (Saha and Idsø 2016). Thus, it is important that
project managers not only involve stakeholders, but also address their concerns (Tabi and
Wüstenhagen 2017).

Energy policy, related to state subsidy and ownership of hydropower, also affects public acceptance. For example, a Greek study found that public acceptance increased when
the state promoted and subsidised hydropower (Ntanos et al. 2018). Regarding ownership, studies from Switzerland found that locals preferred local or state ownership,
then a private domestic company to a foreign investor (Tabi and Wüstenhagen 2017).
The acceptance of hydropower may also be determined by views about other types of
energy sources. Many studies of public acceptance compare public attitudes about different renewable energies (Ribeiro et al. 2014; Schumacher et al. 2019) and found, for
example, that solar and wind were preferred to hydropower and biomass (Botelho et al.
2016).

3.3   Methods of Measuring Public Acceptance

There are different methods of assessing public acceptance of hydropower and its
related technologies. In this sub-section, possible methods are discussed, including the
Q-methodology. The Q-methodology combines the strengths of both qualitative and
quantitative methods. It can identify points of controversy and consensus among different stakeholders as well as compare public perceptions toward hydropower (or other
renewable sources) across regions. Beyond its potential for comparative analysis, the Q-
methodology relies on a small, pre-selected sample of respondents, which can simplify
the survey process. However, preparation of the survey and in-depth interviews can be
laborious.

3.3.1   Comparing Quantitative and Qualitative Methods

Empirical research on public acceptance of renewable energy technologies primarily
employs two types of methodologies: (i) quantitative research methodologies or (ii) qualitative case studies (Devine-Wright 2009). Figure 3.1 compares types of methods for
studying public acceptance.

Quantitative research methods seek to answer questions like "how much" and "how
many". They include large-scale closed surveys, often employing Likert scales or discrete
choice experiments, which elicit an individual's preferences for hypothetical scenarios
or goods (Baxter et al. 2013; Jacquet 2012; Johansson and Laike 2007; Ladenburg and
Dubgaard 2007; Swofford and Slattery 2010). The study of quantitative data primarily
relies on regression analysis. For example, probit and logit models have been used to

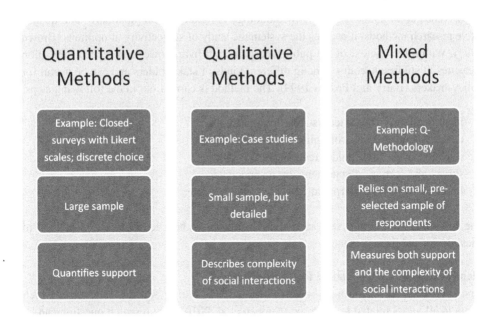

Fig. 3.1 Comparison of methods for studying public perceptions of renewable energy technologies

understand how support for renewable facilities (i.e., large-scale solar, wind) is correlated to demographic variables (gender, age, race, education), socio-psychological measures (i.e., party identification, belief in climate change, annoyance) and geographical variables (postal code, region, size of the city, proximity) (Carlisle et al. 2016; Ladenburg 2008). However, it can be difficult to collect large sample sizes. Further, quantitative studies mainly describe public views at a single point in time (unless panel data is available) rather than explaining the underlying causes of support of or opposition to renewable sources (Devine-Wright 2009).

On the other hand, qualitative research includes analytical induction or grounded theory, often using expert interviews, focus groups or case studies to delve into the unique issues. In such studies, the researcher focuses on studying the complexity of natural human interactions. While qualitative studies allow for the study of contextual issues, they can be time-consuming and labour intensive (Burnard et al. 2008).

3.3.2 Using the Q-Methodology to Study Public Acceptance of Hydropower

There are a growing number of applications of the Q-method to study public views on renewable energy topics, especially for hydropower (Díaz et al. 2017; Pagnussatt et al.

2018; Venus et al. 2020). As the Q-methodology is rooted in both qualitative and quantitative research methods, it enables the systematic study of subjectivity or opinions (Brown 1993). Within the context of the public acceptance of hydropower and related mitigation measures, it studies discourse among different kinds of stakeholders and can be useful for policy-makers (Barry and Proops 1999). The method is carried out in the following steps:

i.   identification of the concourse/research question,
ii. development of the list of opinion statements (Q-set),
iii. selection of the stakeholder respondents (P-set),
iv. survey and sorting of statements by respondents,
v.   factor analysis and interpretation.

The following sub-section is a hands-on description of how to apply the Q-method to questions of public acceptance.

Identification of the Concourse/Research Question

First, the researcher identifies the research question, setting and the concourse, which refers to all views related to the topic (Cuppen et al. 2010). The research question can be phrased as an open question (e.g., "what do locals think about run-of-river hydropower in hydropower-rich regions across Europe?"). The setting should help to define the kind of stakeholders of interest (e.g., within a certain geographic area).

Development of the List of Opinion Statements (Q-Set)

In the second step, the researcher collects a list of opinion statements (Q-set). Statements can be collected directly from stakeholders in interviews or indirectly from the scientific literature or media sources (Brown 1993). In many Q-studies, researchers compile a large set of statements and organize them by similar topics. Then, they iteratively reduce the number of statements. As much as possible, statements should use the original phrasing (inductive). It is also crucial that each statement covers only one topic and reflects an opinion, rather than a fact (Watts and Stenner 2005). Recent applications have also used pictures, rather than written statements (Naspetti et al. 2016).

Researchers should include a sufficient number of statements to cover all opinions on the topic, but also a manageable amount so as not to lead to respondent fatigue. Further, the number of statements (Q-set) should not exceed the number of respondents (P-set). Before implementation, the Q-set should be piloted by testing the Q-sort with experts on the topic or through a validation questionnaire with stakeholder workshops as conducted in Venus et al. (2020).

Selection of the Stakeholder Respondents (P-Set)

As the Q-methodology relies on a small sample size, thoughtful selection of the respondents (P-set) is key. Researchers can make a stakeholder matrix with a minimum and maximum desirable number of respondents per type to guide their selection of respondents. Example types of stakeholders include local authorities, environmental partners, regulators, investors, operators, and local residents. It may also be possible to focus on one type of stakeholder group, provided that differences are made between that type of stakeholder (e.g., operators with different sizes of plants or local residents of different ages). It is also possible to select a geographic area using spatial suitability analysis and survey stakeholders within its borders (Venus et al. 2021).

Survey and Sorting of Statements by Respondents (Q-Sort)

Many Q-studies are carried out in three phases: (i) entry interview, (ii) Q-sort and (iii) exit interview. In the entry interview, the researcher collects information about the respondent. During the Q-sort, the respondent reviews all statements and sorts them into a matrix according to their view on the statement. The matrix reflects a relative ranking with a prompt (Fig. 3.2). For example, the respondent ranks according to how much they agree or disagree with each statement. To prevent cognitive overload, the sorting exercise can have two phases. The respondent could first read all statements and allocate them to three categories: agree, disagree, neutral and then in a second step, distribute them to the grid.

The survey can be conducted on the computer, phone, face-to-face or using a combination. In comparisons of Q-studies by mail, computer-based alternatives and in-person, results were consistent (Exel and Graaf 2005). If using a combination, researchers might use phone entry and exit interviews and an online platform to conduct the Q-sort.

Factor Analysis and Interpretation

Primarily (centroid) factor analysis or principal component analysis were used to analyse the ranking of the statements. This analysis reduces the number of variables (Webler et al. 2009). Researchers should use the eigenvalues, explained variance, number of Q-sorts loading significantly on a component (factor) and theory to determine how many components (factors) to extract (Dziopa and Ahern 2011). The analysis is useful for identifying consensus and controversy across perspectives and stakeholders.

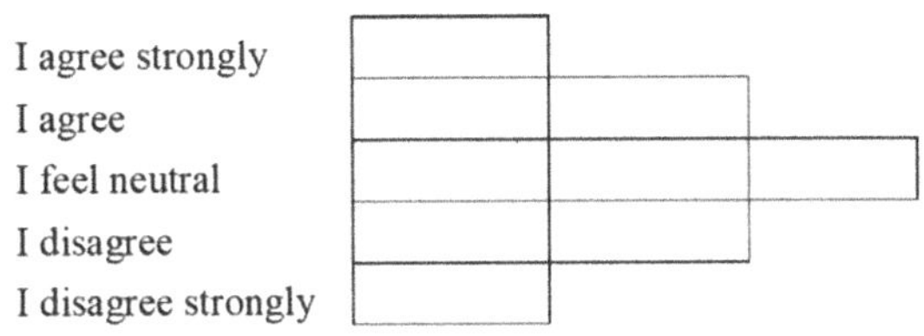

Fig. 3.2 Example Q-sort matrix

I agree strongly
I agree
I feel neutral
I disagree
I disagree strongly

3.4   Example Q Studies of Hydropower Across Europe

3.4.1   Case Studies

To understand how hydropower is perceived by locals across hydropower regions in Europe, Q-studies were conducted in and around Toulouse (France), Landshut (Germany), Vila Real (Portugal) and Örnsköldsvik (Sweden). The locations were selected based on their location in a hydropower-rich region with a nearby hydropower plant and within a 15 km radius of an urban area. To identify similar hydropower contexts, we focused on run-of-river hydropower plants that were less than 20 MW in capacity and had a mitigation measure in place (e.g. fish ladder). The interview sites were selected based on the size of the town (population less than 100,000 people) and proximity to interview teams. Interviews were conducted face-to-face with an interactive poster Q-board.

3.4.2   Results and Discussion

Using principal component analysis, components were extracted. Each component represents a similar opinion pattern or perspective. The components were interpreted based on the rankings of the Q-sets and qualitative data. A more detailed overview of these individual results can be found in Hinzmann et al. (2019) and the combined analysis in Venus et al. (2020). In **Toulouse, France** (n = 46), three components (perspectives) accounted for 49% of total variance: (i) fight climate change, (ii) promote local well-being and (iii) promote fishfriendly and locally owned hydropower. In **Landshut, Germany** (n = 59), three components accounted for 52% of the total variance: (i) promote sustainable energy policy, (ii) preserve rivers, fight climate change and keep it local, (iii) fish protection first. In **Örnsköldsvik, Sweden** (n = 65), three components accounted for 46% of the total variance: (i) fight climate change and create local well-being, (ii) promote regional ownership and (iii) protect habitats and ecosystems. In **Vila Real, Portugal** (n = 87), three components accounted for 40% of the variance: (i) fight climate change and create local well-being, (ii) promote regional ownership and modernization and (iii) protect habitats and ecosystems.

The results demonstrate that there are diverse views among locals. Several important themes emerged: climate protection, ecological effects, local benefits and ownership. In all regions, there was a perspective that supported hydropower because it helps fight climate change. For this group, hydropower was seen as key due to its flexibility and energy storage potential. Further, there was a group in all regions concerned with the negative ecological impacts of hydropower including its effects on fish, habitats and the river ecosystem. The view that hydropower has negative ecological effects was the main reason for opposition to hydropower. Another common pattern across regions was the view that hydropower should benefit locals in the form of job creation, low electricity prices and flood protection. This view was linked to concerns about ownership, which was found in all regions. Respondents who took this view believed that hydropower plants should be owned by companies based in the country, often preferring the state to private owners. They expressed concern that the state may lose its influence over water resources and that foreign/transnational companies may be too focused on profits, leading them to neglect local well-being.

In the context of hydropower mitigation, the results illustrate that a variety of factors are relevant for public support. While many respondents indicated that ecological considerations were key, there was low awareness of mitigation and some expressed doubt regarding its efficacy. While structural measures are likely to be accepted by the public, it is important to provide information about how they function. Morphological measures are most easily observed, thus they are likely to be received positively given that they make rivers appear more "natural". In comparison, operational strategies that support energy storage and increase system flexibility are often unobservable. Nevertheless, they are likely to be perceived positively as long as changes in the river are unobservable (e.g., water levels remain relatively constant). Thus, it is important to improve communication with the public about the spectrum of mitigation measures and their effects on river ecosystems. Operators may also garner public support by highlighting local ownership or green electricity tariffs (Venus et al. 2020).

3.5 Conclusion

Public perceptions can play a decisive role in future hydropower development and modernisation. Particularly during the planning stages of new construction, modernisation or implementation of mitigation measures, it can be valuable to understand and address public and stakeholder views. With this understanding, hydropower managers can better address concerns among the public and various stakeholders. This sub-chapter discussed important factors for public acceptance of hydropower and illustrated how the Q-methodology can be useful for comparing public views about renewable technologies. Compared to purely quantitative surveys, the mixed-method approach of the Q-methodology allows the researcher to identify underlying reasons for public acceptance. In terms of sample size, it requires a comparatively small, pre-selected sample. It is important to note that the method can work with a much smaller sample than demonstrated here. For these reasons, we recommend that practitioners apply the Q-methodology for public acceptances studies.

References

Bakken TH, Sundt H, Ruud A, Harby A (2012) Development of small versus large hydropower in Norway—comparison of environmental impacts. Energy Procedia 20:185–199. https://doi.org/10.1016/j.egypro.2012.03.019

Barry J, Proops J (1999) Seeking sustainability discourses with Q methodology. Ecol Econ 28(3):337–345. https://doi.org/10.1016/S0921-8009(98)00053-6

Baxter J, Morzaria R, Hirsch R (2013) A case-control study of support/opposition to wind turbines: perceptions of health risk, economic benefits, and community conflict. Energy Policy 61:931–943. https://doi.org/10.1016/j.enpol.2013.06.050

Botelho A, Pinto LMC, Lourenço-Gomes L, Valente M, Sousa S (2016) Public perceptions of environmental friendliness of renewable energy power plants. Energy Proc 106:73–86. https://doi.org/10.1016/j.egypro.2016.12.106

Brown S (1993) A primer in Q methodology. Oper Subject 16(3/4):91–138

Burnard P, Gill P, Stewart K, Treasure E, Chadwick B (2008) Analysing and presenting qualitative data. Br Dent J 204(8):429–432. https://doi.org/10.1038/sj.bdj.2008.292

Carlisle JE, Solan D, Kane SL, Joe J (2016) Utility-scale solar and public attitudes toward siting: a critical examination of proximity. Land Use Policy 58:491–501. https://doi.org/10.1016/j.landusepol.2016.08.006

Cuppen E, Breukers S, Hisschemöller M, Bergsma E (2010) Q methodology to select participants for a stakeholder dialogue on energy options from biomass in the Netherlands. Ecol Econ 69(3):579–591. https://doi.org/10.1016/j.ecolecon.2009.09.005

Devine-Wright P (2009) Reconsidering public acceptance of renewable energy technologies: a critical review. In Delivering a Low Carbon Electricity System

Díaz P, Adler C, Patt A (2017) Do stakeholders' perspectives on renewable energy infrastructure pose a risk to energy policy implementation? A case of a hydropower plant in Switzerland. Energy Policy 108:21–28. https://doi.org/10.1016/j.enpol.2017.05.033

Dziopa F, Ahern K (2011) A systematic literature review of the applications of Q-technique and its methodology. Methodology 7(2):39–55. https://doi.org/10.1027/1614-2241/a000021

Exel JV, de Graaf G (2005) Q methodology: a sneak preview. Soc Sci 2:1–30

Gullberg AT, Ohlhorst D, Schreurs M (2014) Towards a low carbon energy future—renewable energy cooperation between Germany and Norway. Renew Energy 68:216–222. https://doi.org/10.1016/j.renene.2014.02.001

Hinzmann M, Gerdes H, Venus T et al (2019) D5.3—Public acceptance of alternative hydropower solutions. FIThydro Project Report. https://www.fithydro.eu/deliverables-tech/

Jacquet JB (2012) Landowner attitudes toward natural gas and wind farm development in northern Pennsylvania. Energy Policy 50:677–688. https://doi.org/10.1016/j.enpol.2012.08.011

Johansson M, Laike T (2007) Intention to respond to local wind turbines: the role of attitudes and visual perception. Wind Energy 10(5):435–451. https://doi.org/10.1002/we.232

Karlstrøm H, Ryghaug M (2014) Public attitudes towards renewable energy technologies in Norway. the role of party preferences. Energy Policy 67:656–663. https://doi.org/10.1016/j.enpol.2013.11.049

Klinglmair A, Bliem M, Brouwer R (2015) Exploring the public value of increased hydropower use: a choice experiment study for Austria. J Environ Econ Policy 4(3):315–336. https://doi.org/10.1080/21606544.2015.1018956

Ladenburg J (2008) Attitudes towards on-land and offshore wind power development in Denmark; choice of development strategy. Renew Energy 33(1):111–118. https://doi.org/10.1016/j.renene.2007.01.011

Ladenburg J, Dubgaard A (2007) Willingness to pay for reduced visual disamenities from offshore wind farms in Denmark. Energy Policy. https://doi.org/10.1016/j.enpol.2007.01.023

Lindström A, Ruud A (2017) Swedish hydropower and the EU Water Framework Directive

Malesios C, Arabatzis G (2010) Small hydropower stations in Greece: the local people's attitudes in a mountainous prefecture. Renew Sustain Energy Rev 14(9):2492–2510. https://doi.org/10.1016/j.rser.2010.07.063

Mattmann M, Logar I, Brouwer R (2016) Hydropower externalities: a meta-analysis. Energy Econ 57:66–77. https://doi.org/10.1016/j.eneco.2016.04.016

Naspetti S, Mandolesi S, Zanoli R (2016) Using visual Q sorting to determine the impact of photovoltaic applications on the landscape. Land Use Policy. https://doi.org/10.1016/j.landusepol.2016.06.021

Ntanos S, Kyriakopoulos G, Chalikias M, Arabatzis G, Skordoulis M (2018) Public perceptions and willingness to pay for renewable energy: a case study from Greece. Sustainability (switzerland) 10(3):687–687. https://doi.org/10.3390/su10030687

Öhman MB, Palo M, Thunqvist EL (2016) Public participation, Human Security and Public Safety around Dams in Sweden: a case study of the regulated Ume and Lule Rivers. Safety Sci Monitor 19(2)

Pagnussatt D, Petrini M, dos Santos ACMZ, da Silveira LM (2018) What do local stakeholders think about the impacts of small hydroelectric plants? Using Q methodology to understand different perspectives. Energy Policy 112:372–380. https://doi.org/10.1016/j.enpol.2017.10.029

Ribeiro F, Ferreira P, Araújo M, Braga AC (2014) Public opinion on renewable energy technologies in Portugal. Energy 69:39–50. https://doi.org/10.1016/j.energy.2013.10.074

Saha P, Idsø J (2016) New hydropower development in Norway: municipalities' attitude, involvement and perceived barriers. Renew Sustain Energy Rev 61:235–244. https://doi.org/10.1016/j.rser.2016.03.050

Schumacher K, Krones F, McKenna R, Schultmann F (2019) Public acceptance of renewable energies and energy autonomy: a comparative study in the French, German and Swiss Upper Rhine region. Energy Policy 126(1):315–332. https://doi.org/10.1016/j.enpol.2018.11.032

Swofford J, Slattery M (2010) Public attitudes of wind energy in Texas: local communities in close proximity to wind farms and their effect on decision-making. Energy Policy 38(5):2508–2519. https://doi.org/10.1016/j.enpol.2009.12.046

Tabi A, Wüstenhagen R (2017) Keep it local and fish-friendly: social acceptance of hydropower projects in Switzerland. Renew Sustain Energy Rev 68:763–773. https://doi.org/10.1016/j.rser.2016.10.006

Venus TE, Hinzmann M, Bakken TH, Gerdes H, Godinho FH, Hansen B, Pinheiro A, Sauer J (2020) The public's perception of run-of-the-river hydropower across Europe. Energy Policy 140. https://doi.org/10.1016/j.enpol.2020.111422

Venus TE, Sauer J (2022) Certainty pays off: the public's value of environmental monitoring. Ecol Econ 191. https://doi.org/10.1016/j.ecolecon.2021.107220

Venus TE, Strauss F, Venus TJ, Sauer J (2021) Understanding stakeholder preferences for future biogas development in Germany. Land Use Policy 109. https://doi.org/10.1016/j.landusepol.2021.105704

Volken S, Wong-Parodi G, Trutnevyte E (2019) Public awareness and perception of environmental, health and safety risks to electricity generation: an explorative interview study in Switzerland. J Risk Res 22(4):432–447. https://doi.org/10.1080/13669877.2017.1391320

Watts S, Stenner P (2005) Doing Q Methodology: theory, method and interpretation. Qual Res Psychol 2(1):67–91. https://doi.org/10.1191/1478088705qp022oa
Webler T, Danielson S, Tuler S (2009) Using Q method to reveal social perspectives in environmental research. Soc Environ Res
Wüstenhagen R, Wolsink M, Bürer MJ (2007) Social acceptance of renewable energy innovation: an introduction to the concept. Energy Policy 35(5):2683–2691. https://doi.org/10.1016/j.enpol.2006.12.001

Impacts and Risks of Hydropower

4

Ruben van Treeck⊙, Juergen Geist⊙, Joachim Pander, Jeffrey Tuhtan, and Christian Wolter⊙

4.1   Introduction

The detrimental effects hydropower plants have on aquatic ecosystems and biodiversity are manifold and comprehensively reviewed (e.g., Gasparatos et al. 2017, Hecht et al. 2019, Jungwirth et al. 2003. Lees et al. 2016, Reid et al. 2019, Schmutz and Sendzimir 2018, Stendera et al. 2012, Ziv et al. 2012). In the following section, however, we review, categorize and outline hydropower-related impacts on freshwater fishes only. This is due

R. van Treeck (✉) · C. Wolter
Leibniz Institute of Freshwater Ecology and Inland Fisheries, Berlin, Germany
e-mail: ruben.vantreeck@ifb-potsdam.de

C. Wolter
e-mail: christian.wolter@igb-berlin.de

R. van Treeck
Ecology of Fish and Waterbodies, Institute of Inland Fisheries Potsdam-Sacrow, Potsdam, Germany

J. Geist · J. Pander
Aquatic Systems Biology, Technical University of Munich, Freising, Germany
e-mail: geist@tum.de

J. Pander
e-mail: joachim.pander@tum.de

J. Tuhtan
Department of Computer Systems, School of Information Technologies, Tallinn University of Technology, Tallinn, Estland
e-mail: jeffrey.tuhtan@taltech.ee

© The Author(s) 2022
P. Rutschmann et al. (eds.), *Novel Developments for Sustainable Hydropower*,
https://doi.org/10.1007/978-3-030-99138-8_4

41

to various reasons: For one, fishes are of great socio-economic interest. Their unquestionable cultural and societal value has caused managing efforts to support self-sustained, exploitable fish stocks for several thousand years, and today they are a priority target for many restoration and conservation programs. Furthermore, fish are most affected by the operation of hydropower (Larinier 2001) and the high level of hydromorphological degradation and resulting habitat loss associated with hydropower has been identified as one of the bottlenecks in reaching the Water Framework Directive targets (Freyhof et al. 2019).

Therefore, this chapter draws a comprehensive conceptual model depicting what kinds of impacts on fish potentially happen beginning from habitat loss/modification upstream due to the impoundment, migration delays, indirect mortality due to increased predation, the hydropower plant (HPP) itself, with potential spillway, bypass, trash racks and also turbine effects (blade strike, shear forces, barotrauma) and down to tailwater effects, such as increased predation, residual flows, habitat and flow modifications (Fig. 4.1).

The resilience of fish species and populations as well as species most at risks will be addressed based on narratives derived as risk factors and the empirical evidence provided by the literature review.

4.2   Barrier Effects

The central, most prominent element of every hydropower scheme is undoubtedly a dam or a weir. Although these types of barriers are not exclusive to hydropower plants, they always have the same principal effects on fishes. Because barriers become impassable

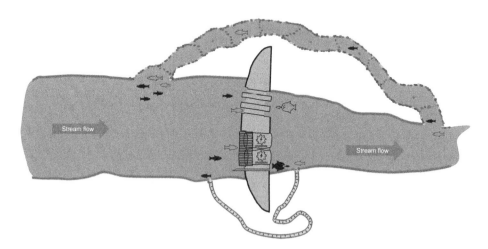

Fig. 4.1 Conceptual sketch of elements of a hydropower scheme potentially impacting fish, like the barrier itself, upstream and downstream migration routes, turbines, trash racks and fish protection facilities

obstacles for fishes once they exceed certain dimensions, they segregate resident populations into isolated upstream and downstream components. Barriers disrupt the original river continuum (Allan and Castillo 2007; Mueller et al. 2011; Vannote et al. 1980) and the natural migration corridors for fishes (Jonsson et al. 1999). Dams and weirs act as migration barriers for migratory species that then face substantial migration delay (Buysse et al. 2015; Ovidio et al. 2017; Stich et al. 2015, Winter et al. 2006), and they render critical habitats inaccessible to fishes (Larinier 2001; Pelicice et al. 2015). However, unhindered upstream migration is particularly critical for diadromous migratory species like salmonids, lampreys, some clupeids or sturgeons that only spawn in the upper regions of rivers where hydraulic and geomorphic conditions support egg development and provide larval habitats (Katano et al. 2006; Lucas et al. 2009; Penczak et al. 1998). But also, migrations of potamodromous species are impaired by barriers (Britton and Pegg 2011; Lucas and Frear 1997). This can result in reduced natural recruitment (McCarthy et al. 2008), differences in population structure and species assemblages up- and downstream of the dam (Franssen and Tobler 2013; Morita and Yamamoto 2002; Mueller et al. 2011) and even result in the extinction of entire fish stocks (Larinier and Travade 1992), unless habitat heterogeneity and availability in the system remains high enough to support the native assemblage (Santos et al. 2006). Furthermore, because dams act as bi-directional nutrient traps that can cause a reduction of far-downstream fish biomass (Jackson and Marmulla 2001) and a lack of nutrients (i.e., due to a lower number of spawners remaining in the headwaters of streams), which directly affects the dietary composition of a range of different fish species (Piorkowski 1995). The mechanisms described in this paragraph primarily impact population endpoints that ultimately, cause a decline in recruitment, whereas individual mortality of affected fishes is only of secondary concern.

The negative ecological impacts of barriers can be partly mitigated by maintaining certain flow velocity through the impounded area that resembles the ecological functioning of the former stream. These flow patterns are cues for up- and downstream migrating species and ensure sediment transport and aeration.

4.3   Upstream Flow Alterations

Dams cause substantial alterations of the stream's original discharge regime (Egré and Milewski 2002; Schiemer et al. 2001). Reservoirs and impoundments considerably slow down the stream's flow velocity causing higher sedimentation rates of finer particles, stratification, increased temperature, and potential oxygen depletion in the hypolimnion due to an imbalance in aerobic production and consumption (Thornton et al. 1990). In principle, impoundments transform lotic habitats into ones with more lentic characteristics (Sá-Oliveira et al. 2015) that are unsuited for most riverine, lithophilic species that require well aerated, fast flowing coarse gravel beds as spawning habitats (Wood and Armitage 1997). These conditions result in habitat loss for a range of rheophilic species (Agostinho et al. 2002; Birnie-Gauvin et al. 2017; Larinier 2001; Tiffan et al. 2016), changes in water quality (Fantin-Cruz et al. 2016), shifts in biomass and ultimately, changes of species abundance and diversity relative to non-impounded reaches downstream (Sá-Oliveira et al. 2015). These conditions also affect species-specific length-frequency distributions, species richness (Gehrke et al. 2002) and species composition (Tundisi and Straškraba 1999). Manipulated abiotic conditions in impoundments were further associated with temperature-related changes of growth patterns (i.e., younger age of maturity and smaller individual sizes) (Reed et al. 1992). For example, another study by Yang et al. (2020) showed reduced energy transfer efficiency in impoundments, suggesting potential energetic bottlenecks of fish at higher trophic levels. In impoundments altered hydromorphological conditions have caused increased predation, most likely because of the novel environment, lack of navigation cues for diadromous species (Agostinho et al. 2002; Jepsen et al. 2000; Tiffan et al. 2016) and the resulting migration delay (Larinier 2001; Larinier and Travade 2002) and reinforce negative impacts of introduced predators (Pelicice and Agostinho 2009). This can lead to local extinction of native and proliferation of non-native species (Martinez et al. 1994).

4.4   Downstream Flow Alterations

Different types of HPP have to be distinguished. There are run-of-river HPP of both instream or diversion-type schemes and storage HPP as well as pump-storage plants (Egré and Milewski 2002; Matt et al. 2019). Particularly storage, but to some extent also run-of-river hydropower plants dampen high natural discharge amplitudes by cutting flow peaks and increasing very low discharges. As such, they completely alienate the natural discharge regime of a stream, with flow fluctuations downstream being most problematic at all plants that do not release approximately as much water through the dam (i.e., through the turbines, spill gates or sluices) as would normally be discharged in the stream.

In diversion plants, the main purpose of the dam is to divert water away from the main stream towards the (potentially very remote) powerhouse where the water is turbinated and returned to the original river bed further downstream (Egré and Milewski 2002). The residual old river bed usually suffers from water scarcity, and methodological frameworks for defining sufficient environmental flow in the affected stretch are summarized by the CIS Guidance 31 "Ecological Flows in the Implementation of the Water Framework Directive" that can be consulted to mitigate the negative effects. At HPPs in which only a fraction of the original discharge remains in the residual river stretch severe consequences regarding water depths, flow velocities, and temperature extremes were observed. These do not support some fish populations anymore, cause species shifts and population declines (Anderson et al. 2006; Benejam et al. 2016; Habit et al. 2007) and sometimes even render whole river stretches uninhabitable. At some HPPs with state-of-the-art environmental flows of at least 10% mean annual stream flow (Huckstorf et al. 2008) these impacts are less pronounced. However, maintaining the comparably high environmental flow usually comes at the expense of hydroelectricity generation and loss of revenues.

Hydropeaking plants typically store larger amounts of water and release it for electricity generation in times of peak demand, mostly in the morning and evening (Moreira et al. 2019; Schmutz et al. 2015; Schmutz and Sendzimir 2018). Many species cannot cope with manipulated flow alterations induced by turbine operation which can lead to reduced food availability (De Jalon et al. 1994; Gandini et al. 2014; Young et al. 2011), erosion and habitat loss due to periodical dewatering (Almodóvar and Nicola 1999; Boavida et al. 2015, 2013; Choi et al. 2017; Person 2013; Shen and Diplas 2010; Young et al. 2011) and impaired egg development (Casas-Mulet et al. 2015a, b; Person 2013; Young et al. 2011), all of which commonly resulting either in reduced recruitment or increased direct mortality e.g., by stranding (Hedger et al. 2018, Schmutz et al. 2015, Tuthan et al. 2012, Young et al. 2011) in particular of smaller species or younger specimen with weaker swimming performance (Hayes et al. 2019; Person 2013).

If water shortage or pulse flows are not evident, manipulated flows can still exert major pressures on fishes e.g., because new habitat types immediately emerge beneath the dam that support accumulation of fishes (Jackson 1985) that attract unnaturally high abundances of predators able to deplete already impaired stocks (Larinier 2001; Stansell et al. 2010). In addition, hydropeaking can lead to altered sediment dynamics in rivers with severe consequences for lithophilic fish species (Casas-Mulet et al. 2015a, b).

4.5   Upstream Passage

Upstream migration needs of fishes have received much more attention relative to downstream migration needs, and respective efforts to increase passage rates date back longer, too (Katopodis and Williams 2012). The decline of the highly valued anadromous salmonids and the respective fisheries in response to damming became obvious very early on and had resulted in first legal acts that obliged e.g., mill owners to care about fish migration. In this context, attempts to facilitate upstream movement of fish that actively search for passage corridors have been more successful compared to attempts to guide fish following the main current in downstream direction (Geist 2021). Correspondingly, comprehensive guidelines exist to facilitate operational upstream migration facilities under varying environmental, technical and biotic conditions e.g., the DWA guidance M 509 (DWA 2014). However, upstream migration facilities show highly varying passage rates between 0 and 100% (Bunt et al. 2012; Gowans et al. 2003; Hershey 2021; Kemp et al. 2011), mostly due to the unique and highly complex interaction between the species' internal state and motivation to migrate, their anatomy and swimming ability, ambient hydraulic conditions and type and design of the passage facility (Banks 1969, Castro-Santos et al., 2009, Crisp 2000, USFWS (U.S. Fish and Wildlife Service) 2019). In Europe

the implementation of the WFD stipulated the re-establishment of the longitudinal connectivity (Schletterer et al. 2016) and various technical as well as natural fishways were
developed or species-specifically improved (Clay 2017; Hershey 2021; Jungwirth et al.
1998; Katopodis 1992; Santos et al. 2014).

Factors determining passage success of an upstream fishpass include attraction efficiency mediated by position of entrance and attraction flow and passability mediated by
slope, flow velocity in the migration corridor, height differences and physical dimensions (Banks 1969; Bunt et al. 2012; DWA 2014; Hershey 2021, USFWS 2019). Failing
upstream passage success of fish result in excessive energy expenditure and migration
delays (Noonan et al. 2012; Silva et al. 2019; Thorstad et al. 2008) and thus, delayed
arrival at spawning events (Silva et al. 2019), and increased predation (Agostinho et al.
2012). When HPPs are aligned in cascades their cumulative barrier effects must be considered (Geist 2021) as it aggravates already significant delays, migration failures and
mortalities threatening the persistence of fish populations (Caudill et al. 2007; Gowans
et al. 2003; Muir et al. 2001; Roscoe et al. 2011; Williams et al. 2001).

4.6   Downstream Passage

Downstream passage attained attention only much more recently, but is of similar relevance especially for iteroparous species spawning more than once in a lifetime. Beside
the target species (diadromous or potamodromous) and the biocoenotic region (upper vs.
lower course and associated species guilds) also HPP constellation (size, turbine type,
etc.) and operational mode need to be considered (Schmidt et al. 2018; Travade and Larinier 2002). Particularly, juveniles of anadromous and adults of catadromous guilds but
also potamodromous species require unobstructed downstream migration corridors. Therefore, HPPs must be equipped with fish guiding structures that facilitate downstream fish
migration. Generally, all routes downstream over barriers and through HPPs are inherently
dangerous for fishes and may result in migration delay or elevated mortality.

Spillways, mostly used to release excess water in times of higher discharge, can serve
as effective and comparably fish-friendly downstream paths through a hydropower plant
with bypass efficiencies of >90% (Muir et al. 2001). However, water released through
spillways, particularly from bigger heights, tends to supersaturate with nitrogen and oxygen and, together with shear forces, pressure changes and blunt trauma or abrasions, can
cause substantial damages and high mortality rates: up to 2% at a height of <3 m, up to
40% at 10 m and up to 100% at 50 m (Algera et al. 2020; Heisey et al. 1996; Schilt 2007;
Wolter et al. 2020), with larger fish being significantly more susceptible to drop-induced
injuries than smaller ones (Ruggles and Murray 1983).

Sluice gates installed at hydropower plants are mostly opened to spill debris or discharge excess inflow and may constitute temporarily available pathways for downstream
migrating fish, too. Because the hydraulic conditions around an open (esp. undershot)

gate act as a strong cue for migrating species sluices have proven efficient in conveying e.g., European eels downstream (Egg et al. 2017). However, undershot pathways may expose passing fish to rapid pressure changes that by far exceed those at overshot routes (Pflugrath et al. 2019), causing up to 95% mortality rates, especially for juveniles, small species and those with pressure-sensitive swim bladders (Algera et al. 2020; Baumgartner et al. 2006; Marttin and De Graaf 2002), while passage efficiency varies between <20% (Kemp et al. 2011) and >90% (Gardner et al. 2016).

Bypasses are dedicated downstream migration routes for fishes and most often used in combination with deflection screens or behavioural guidance facilities (Ebel et al. 2015). Their set-up is usually relatively simple, comprising concrete or metal chutes, slides or pipes that flush entering fishes downstream. Operational and efficient bypasses must be easily accessible, sufficiently dimensioned and supplied with enough water (commonly measured as a proportion of the turbine flow rate), and the entering water should have a slightly higher flow velocity than the recommended approaching flow of deflection screens (Ebel et al. 2015; Larinier and Travade 2002). Studies quantifying bypass mortalities are comparably scarce (Algera et al. 2020), but documented bypass-related damages and mortalities are mainly caused by sheer forces, rapid pressure changes, collisions, disorientation and subsequent predation in the tailrace (Williams et al. 2001); however, mortalities remained generally lower compared to other downstream routes (Algera et al. 2020). Bypass passage rates of fish showed significant variation between 0 and 95% (Gosset et al. 2005; Nyqvist et al. 2018; Ovidio et al. 2017).

Trash racks are installed in front of turbine intakes to protect them from large debris like wood. Normally, they feature vertical bars that—depending on design requirements— may be slightly inclined. The bar spacing is usually very wide to minimize head loss and constitute a substantial risk for larger fish that may get impinged and damaged when the approaching flow velocity is too high, during trash cleaner operations or when debris accumulates in the forebay (Weibel 1991). Studies investigating mortality rates of fishes due to trash racks are methodologically very challenging and thus, scarce.

Deflection screens with much smaller bar spacing installed at HPP behind or instead of trash racks are mechanical and behavioural barriers that prevent fishes from entering the turbines. Fish deflection screens come in a wide variety of designs e.g., vertically inclined with vertical bars and horizontally angled screens with horizontal bars that mostly deflect fishes mechanically, or horizontally angled screens with vertical bars inducing an additional behavioural change that increases the deflection performance up to 95% (Albayrak et al. 2020; Amaral 2003; Beck 2019; Calles et al. 2013; Ebel 2013a; Ebel et al. 2015; Nyqvist et al. 2018). The purely mechanical deflection rate can be approximated using empirical length-width-regressions by (Ebel 2013b): for example, 18 mm bar spacing would deflect fusiform fish of approximately ≥16 cm and eel of approximately ≥55 cm length; 15 mm bar spacing would lower these values to 13.6 and 48 cm. In contrast, a common trash rack with 80–100 mm bar spacing is consequently passable for almost all native species. When the approaching flow exceeds the recommended value of approximately 0.5 m/s (Calles et al. 2013; DWA 2014; Ebel et al. 2015; Larinier and Travade 2002), fish may be impinged in the screen and get damaged (Calles et al. 2013; Larinier 2001). Typically, physical/behavioural deflection screens and downstream bypasses form a functional unit (Ebel et al. 2015; Gosset et al. 2005; Larinier and Travade 2002; Nyqvist et al. 2018; Økland et al. 2019) and are not considered operational in absence of each other.

Turbine passage is probably the best-studied, most dangerous downstream route for fishes (Algera et al. 2020, Eicher et al. 1987). Depending on type and size of the turbine, fishes can get damaged or killed usually by either one or a combination of i) abrupt pressure changes (barotrauma), ii) turbulent flow, iii) shear forces, and iv) turbine blade strikes (USFWS 2019). Generally, the consequences of direct and delayed mortality as well as external (Mueller et al. 2017) and internal (Mueller et al. 2020a, b, c, d, e, f, g, h, i) injuries following turbine passage must be distinguished. Reported mortalities were highly variable across and within turbine types e.g., 1–7.7% in "Very Low Head" (VLH) turbines (Hogan et al. 2014; Reuter and Kohout 2014), 2% in Alden turbines (Hogan et al., 2014), 2–2.4% for the "Minimum Gap Runner" (MGR) (Čada et al. 1997; Hogan et al. 2014), 0.1–2.5% in water wheels (Pulg and Schnell 2008; Quaranta and Wolter 2021; Reuter and Kohout 2014), 0–32.7% in Archimedes screws (Buysse et al. 2015; Hogan et al. 2014; Piper et al. 2018; Pulg and Schnell 2008; Reuter and Kohout 2014), 0.3–100% in Kaplan turbines (Anon et al. 1987, Čada et al. 1997, 2006; Čada 2001; Cramer and Oligher 1964; Reuter and Kohout 2014; Richmond et al. 2014), although the risk of lethal blade strike in large Kaplan turbines can be substantially reduced compared to that of smaller ones (Bell and Kynard 1985), 15 to >70% in Ossberger turbines (Gloss and Wahl 1983), 4–100% in Francis turbines (Anon et al. 1987, Cramer and Oligher 1964; Pulg and Schnell 2008; Reuter and Kohout 2014) and 100% in Pelton wheels (Reuter and Kohout 2014). Fish mortality increases with increasing rotational speed (Anon et al. 1987, Buysse et al. 2015; Cramer and Oligher 1964; Odeh 1999; Turnpenny et al. 2000) usually inversely correlates with turbine size and positively correlates with fish size (Čada 1990; Colotelo et al. 2012; Pracheil et al. 2016) and hydraulic head (Anon et al. 1987, Larinier 2001) i.e., with rapid decompression and lack of acclimation time (Brown et al. 2009, 2012; Colotelo et al. 2012; Cramer and Oligher 1964; Odeh 1999; Pracheil et al. 2016; Richmond et al. 2014; Stephenson et al. 2010; Turnpenny et al. 2000). Further, mortality decreases with increasing turbine load (Čada et al. 1997; Cramer and Oligher 1964) and depends on fish behaviour and species (Amaral et al. 2015; Calles et al. 2010; Coutant and Whitney 2000; Ebel 2013a; Havn et al. 2017). Even if direct mortality rates are not evident, fishes may die from their injuries later (Ferguson et al. 2006; Mueller et al. 2020c, 2020f, 2020a, 2020e, 2020d, 2020b, 2020g; Muir et al. 2006; Taylor and Kynard 1985). This delayed mortality can be substantial and not accounting for it might severely underestimate damage rates during field studies and therefore, must be considered in the experimental design.

Turbine entrainment can cause damages and mortalities, and thus, be a significant population impact factor not only for juveniles with weaker swimming abilities or migratory species (i.e., salmonid smolts) (Mathur et al. 2000; Thorne and Johnson 1993) but also for potamodromous (Harrison et al. 2019) and even resident adult fishes, mainly in fall and winter (Martins et al. 2013). However, survival for smaller (i.e., juvenile) fishes at turbine passage is often higher than for adults, and turbine entrainment may therefore contribute to the persistence of downstream populations, albeit at the expense of populations upstream (Amaral et al. 2018; Harrison et al. 2019). Entrainment and mortality of drifting fish larvae are severely understudied and have not been quantified so far.

4.7 Risk and Impact Assessment

Measuring, describing, and predicting the actual impact of a HPP or specific, hydropower-related stressors on fish populations is challenging and almost impossible, regardless of the knowledge about single, site- or constellation-specific factors. This is due to several reasons.

First, the lack of information on the reference state, that is the undisturbed condition of the system (Nijboer et al. 2004). The fundamental elements of many HPP (i.e., dams or weirs) are fairly old, and (at least in Europe) new, and particularly small hydropower plants are commonly built on top of existing infrastructure. This imposes serious constraints on typical means of impact investigations like BA (before-after) or BACI (before-after-control-impact) designs (Conner et al. 2015b; Eberhardt 1976; Green 1979; Smith 2014), unless the scientific objective is to assess the additional impact or mortality factor of the hydropower plant compared to that of the already existing dam. If construction work on the HPP or dam has not yet started studies applying BACI designs could be used to investigate hydropower-related impacts before and after completion (e.g., Almodóvar and Nicola 1999), but if a particular stressor is already in place meaningful conclusions about its impact are more difficult to obtain. Pressure-release studies, for example in the context of dam removals or restoration (Catalano et al. 2007; Conner et al. 2015a), could identify improvements from the prevalent condition without knowledge about the reference condition. However, such studies merely describe the "opportunistic" response of the ecosystem and not its resilience i.e., its proximity to the pre-disturbance state. Further, most river systems are facing multiple stressors (Mueller et al. 2020a, b, c, d, e, f, g, h, i) and the single impacts of HPPs are hard to disentangle.

Second, investigations of impacts from hydropower on fish populations are biased towards migratory (i.e., diadromous) species that express clearly distinguished, life stage-critical habitat shifts (Geist 2021). Species with a pronounced migration tendency like anadromous salmonids and lampreys will by default always attempt to pass the hydropower plant if their spawning or rearing grounds are located upstream of the plant. In contrast, it becomes much more difficult to detect impacts at the population level of resident, non-migratory or potamodromous species that do not express long-distance migratory behaviour, migrate within the river system or even stay in the impoundment.

Furthermore, the complexity of different hydropower-related stressors, their interactions, cumulative effects on river system scale (Geist 2021) and summed impact on resident or migratory fishes raise difficulties in predicting their impact in isolation, especially in relation to varying susceptibility of fish assemblages across sites. Conclusions drawn from observations at one site are not necessarily valid at another. While the constellation of a few hydropower components (e.g., turbine type and hydraulic head or turbine size, rotational speed and flow rate) will remain relatively constant across sites and applications, others are much more subject to either the operator's intentions (e.g., operation modes), geo- and hydro-morphologically imposed structural design decisions (e.g., plant type, stream and discharge, mode of operation), spatial limitations (e.g., upstream migration facilities), composition and diversity of the ambient fish community, and fish protection facilities installed (e.g., dimensions of fish deflection screens and design or location of bypass systems). These elements can not only be combined in many different ways, they also interact uniquely with fish species and their life stages. Last but not least, site-specific environmental and conservation concerns do not only constrain the implementation details of a HPP, they also frame the environmental impact assessment. In conservation priority areas, even low impacts from hydropower might not be tolerable, while in heavily modified rivers HPPs of moderate impact might be acceptable.

References

Agostinho AA, Agostinho CS, Pelicice FM, Marques EE (2012) Fish ladders: safe fish passage or hotspot for predation? Neotrop Ichthyol 10(4):687–696. https://doi.org/10.1590/S1679-622520 12000400001

Agostinho AA, Gomes LC, Fernandez R, Suzuki HI (2002) Efficiency of fish ladders for neotropical ichthyofauna. 306:299–306. https://doi.org/10.1002/rra.674

Albayrak I, Boes RM, Kriewitz-Byun CR, Peter A, Tullis BP (2020) Fish guidance structures: hydraulic performance and fish guidance efficiencies. J Ecohydraul 1–19. https://doi.org/10.1080/24705357.2019.1677181

Algera DA, Rytwinski T, Taylor JJ, Bennett JR, Smokorowski KE, Harrison PM, Clarke KD, Enders EC, Power M, Bevelhimer MS (2020) What are the relative risks of mortality and injury for fish during downstream passage at hydroelectric dams in temperate regions? Systemat Rev Environ Evid 9(1):3

Allan JD, Castillo MM (2007) Stream ecology: structure and function of running waters. Springer Science & Business Media

Almodóvar A, Nicola GG (1999) Effects of a small hydropower station upon brown trout Salmo trutta L. in the River Hoz Seca (Tagus basin, Spain) one year after regulation. Regulat Rivers Res Manage 15(5):477–484. https://doi.org/10.1002/(SICI)1099-1646(199909/10)15:5<477::AID-RRR560>3.0.CO;2-B

Amaral SV (2003) The use of angled bar racks and louvers for guiding fish at FERC-Licensed projects. In: FERC Fish Passage Workshop, November 13, 2003, 37

Amaral SV, Bevelhimer MS, Čada GF, Giza DJ, Jacobson PT, McMahon BJ, Pracheil BM (2015) Evaluation of behavior and survival of fish exposed to an axial-flow hydrokinetic turbine. North Am J Fish Manage 35(1):97–113. https://doi.org/10.1080/02755947.2014.982333

Amaral SV, Coleman BS, Rackovan JL, Withers K, Mater B (2018) Survival of fish passing downstream at a small hydropower facility. Mar Freshw Res 69(12):1870–1881

Amaral SV, Watson SM, Schneider AD, Rackovan J, Baumgartner A (2020) Improving survival: injury and mortality of fish struck by blades with slanted, blunt leading edges. J Ecohydraul 5(2):175–183. https://doi.org/10.1080/24705357.2020.1768166

Anderson EP, Freeman MC, Pringle CM (2006) Ecological consequences of hydropower development in Central America: impacts of small dams and water diversion on neotropical stream fish assemblages. River Res Appl 22(4):397–411. https://doi.org/10.1002/rra.899

Banks JW (1969) A review of the literature on the upstream migration of adult salmonids. J Fish Biol 1(2):85–136. https://doi.org/10.1111/j.1095-8649.1969.tb03847.x

Baumgartner LJ, Reynoldson N, Gilligan DM (2006) Mortality of larval Murray cod (Maccullochella peelii peelii) and golden perch (Macquaria ambigua) associated with passage through two types of low-head weirs. Mar Freshw Res 57(2):187–191

Beck C (2019) Hydraulic and fish-biological performance of fish guidance structures with curved bars. In: 38th International Association for Hydro-Environmental Engineering and Research World Congress (IAHR 2019)

Bell CE, Kynard B (1985) Mortality of adult American shad passing through a 17-megawatt Kaplan turbine at a low-head hydroelectric dam. North Am J Fish Manage 5(1):33–38. https://doi.org/10.1577/1548-8659(1985)5%3c33:moaasp%3e2.0.co;2

Benejam L, Saura-Mas S, Bardina M, Solà C, Munné A, García-Berthou E (2016) Ecological impacts of small hydropower plants on headwater stream fish: from individual to community effects. Ecol Freshw Fish 25(2):295–306. https://doi.org/10.1111/eff.12210

Birnie-Gauvin K, Aarestrup K, Riis TMO, Jepsen N, Koed A (2017) Shining a light on the loss of rheophilic fish habitat in lowland rivers as a forgotten consequence of barriers, and its implications for management. Aquat Conserv Mar Freshwat Ecosyst 27(6):1345–1349. https://doi.org/10.1002/aqc.2795

Boavida I, Santos JM, Ferreira MT, Pinheiro A, Zhaoyin W, Lee JHW, Jizhang G, Shuyou C (2013). Fish habitat-response to hydropeaking. In: Proceedings of the 35th Iahr World Congress, Vols I and Ii, August 2015, 1–8

Boavida I, Santos JM, Ferreira T, Pinheiro A (2015) Barbel habitat alterations due to hydropeaking. J Hydro Environ Res 9(2):237–247. https://doi.org/10.1016/j.jher.2014.07.009

Britton JR, Pegg J (2011) Ecology of European barbel Barbus barbus: implications for river, fishery, and conservation management. Rev Fish Sci 19(4):321–330. https://doi.org/10.1080/10641262.2011.599886

Brown RS, Carlson TJ, Gingerich AJ, Stephenson JR, Pflugrath BD, Welch AE, Langeslay MJ, Ahmann ML, Johnson RL, Skalski JR, Seaburg AG, Townsend RL (2012) Quantifying mortal injury of juvenile Chinook salmon exposed to simulated hydro-turbine passage. Trans Am Fish Soc 141(1):147–157. https://doi.org/10.1080/00028487.2011.650274

Brown RS, Carlson TJ, Welch AE, Stephenson JR, Abernethy CS, Ebberts BD, Langeslay MJ, Ahmann ML, Feil DH, Skalski JR (2009) Assessment of barotrauma from rapid decompression of depth-acclimated juvenile Chinook salmon bearing radiotelemetry transmitters. Trans Am Fish Soc 138(6):1285–1301

Bunt CM, Castro-Santos T, Haro A (2012) Performance of fish passage structures at upstream barriers to migration. River Res Appl 28(4):457–478. https://doi.org/10.1002/rra.1565

Buysse D, Mouton AM, Baeyens R, Coeck J (2015) Evaluation of downstream migration mitigation actions for eel at an Archimedes screw pump pumping station. Fish Manage Ecol 22(4):286–294. https://doi.org/10.1111/fme.12124

Čada GF (1990) A review of studies relating to the effects of propeller-type turbine passage on fish early life stages. North Am J Fish Manage 10(4):418–426. https://doi.org/10.1577/1548-867 5(1990)010%3c0418:arosrt%3e2.3.co;2

Čada GF (2001) The development of advanced hydroelectric turbines to improve fish passage survival. Fisheries 26(9):14–23. https://doi.org/10.1577/1548-8446(2001)026%3c0014:tdoaht% 3e2.0.co;2

Čada GF, Coutant CC, Whitney RR (1997) Development of biological criteria for the design of advanced hydropower turbines (Issue i). EERE Publication and Product Library, Washington, DC (United States)

Čada G, Loar J, Garrison L, Fisher R, Neitzel D (2006) Efforts to reduce mortality to hydro-electric turbine-passed fish: locating and quantifying damaging shear stresses. Environ Manage 37(6):898–906. https://doi.org/10.1007/s00267-005-0061-1

Calles O, Olsson IC, Comoglio C, Kemp PS, Blunden L, Schmitz M, Greenberg LA (2010) Size-dependent mortality of migratory silver eels at a hydropower plant, and implications for escape-ment to the sea. Freshw Biol 55(10):2167–2180. https://doi.org/10.1111/j.1365-2427.2010.024 59.x

Calles O, Rivinoja P, Greenberg L (2013) A historical perspective on downstream passage at hydro-electric plants in Swedish rivers. Ecohydraulics: an integrated approach. John Wiley & Sons, Ltd, 309–321

Casas-Mulet R, Alfredsen K, Hamududu B, Timalsina NP (2015a) The effects of hydropeaking on hyporheic interactions based on field experiments. Hydrol Process 29(6). https://doi.org/10.1002/ hyp.10264

Casas-Mulet R, Saltveit SJ, Alfredsen K (2015b) The survival of Atlantic salmon (Salmo salar) eggs during dewatering in a river subjected to hydropeaking. River Res Appl 31(4):433–446. https:// doi.org/10.1002/rra.2827

Castro-Santos T, Cotel A (2015) Webb P (2009) fishway evaluations for better bioengineering : an integrative approach a framework for fishway. Am Fish Soc Symp 69:557–575

Catalano MJ, Bozek MA, Pellett TD (2007) Effects of dam removal on fish assemblage structure and spatial distributions in the Baraboo River, Wisconsin. North Am J Fish Manag 27(2):519–530

Caudill CC, Daigle WR, Keefer ML, Boggs CT, Jepson MA, Burke BJ, Zabel RW, Bjornn TC, Peery CA (2007) Slow dam passage in adult Columbia River salmonids associated with unsuccessful migration: delayed negative effects of passage obstacles or condition dependent mortality? Can J Fish Aquat Sci 64(7):979–995. https://doi.org/10.1139/F07-065

Choi SU, Kim SK, Choi B, Kim Y (2017) Impact of hydropeaking on downstream fish habitat at the Goesan Dam in Korea. Ecohydrology 10(6). https://doi.org/10.1002/eco.1861

Clay CH (2017) Design of fishways and other fish facilities. CRC Press, In Design of Fishways and Other Fish Facilities. https://doi.org/10.1201/9781315141046

Colotelo AH, Pflugrath BD, Brown RS, Brauner CJ, Mueller R, Carlson TJ, Deng ZD, Ahmann ML, Trumbo BA (2012) The effect of rapid and sustained decompression on barotrauma in juvenile brook lamprey and Pacific lamprey: implications for passage at hydroelectric facilities. Fish Res 129–130:17–20. https://doi.org/10.1016/j.fishres.2012.06.001

Conner MM, Saunders WC, Bouwes N, Jordan C (2015a) Evaluating impacts using a BACI design, ratios, and a Bayesian approach with a focus on restoration. Environ Monit Assess 188(10). https://doi.org/10.1007/s10661-016-5526-6

Conner MM, Saunders WC, Bouwes N, Jordan C (2015b) Evaluating impacts using a BACI design, ratios, and a Bayesian approach with a focus on restoration. Environ Monit Assess 188(10). https://doi.org/10.1007/s10661-016-5526-6

Coutant CC, Whitney RR (2000) Fish behavior in relation to passage through hydropower turbines: a review. Trans Am Fish Soc 129(2):351–380. https://doi.org/10.1577/1548-8659(2000)129%3c0 351:fbirtp%3e2.0.co;2

Cramer FK, Oligher RC (1964) Passing fish through hydraulic turbines. Trans Am Fish Soc 93(3):243–259. https://doi.org/10.1577/1548-8659(1964)93[243:pftht]2.0.co;2

Crisp DT (2000). Trout and Salmon: Ecology. Conservation and Rehabilitation, Fishing News Books, Blackwell, Oxford.

De Jalon DG, Sanchez P, Camargo JA (1994) Downstream effects of a new hydropower impoundment on macrophyte, macroinvertebrate and fish communities. Regul Rivers: Res Manage 9(4):253–261. https://doi.org/10.1002/rrr.3450090406

DWA (2014). Merkblatt DWA-M 509: Fischaufstiegsanlagen und fischpassierbare Bauwerke. Report: Merkblatt, 27

Ebel G (2013a) Fischschutz und Fischabstieg an Wasserkraftanlagen. Handbuch Rechen-Und Bypasssysteme. Bd, 4

Ebel G (2013b) Fish Protection and Downstream Passage at Hydro Power Stations Handbook of Bar Rack and Bypass Systems: Büro für Gewässerökologie und Fischereibiologie

Ebel G, Gluch A, Kehl M (2015) Einsatz des leitrechen-bypass-systems nach Ebel, Gluch & Kehl an wasserkraftanlagen—Grundlagen Erfahrungen Und Perspektiven. Wasserwirtschaft 105(7–8):49

Eberhardt LL (1976) Quantitative ecology and impact assessment. J Environ Manage (United Kingdom) 4(1)

Egg L, Mueller M, Pander J, Knott J, Geist J (2017) Improving European Silver Eel (Anguilla anguilla) downstream migration by undershot sluice gate management at a small-scale hydropower plant. Ecol Eng 106:349–357. https://doi.org/10.1016/j.ecoleng.2017.05.054

Egré D, Milewski JC (2002) The diversity of hydropower projects. Energy Policy 30(14):1225–1230. https://doi.org/10.1016/S0301-4215(02)00083-6

Eicher GJ, Bell MC, Campbell CJ, Craven RE, Wert MA (1987) Turbine-related fish mortality: review and evaluation of studies. Palo Alto, CA, Electric Power Research Institute, Report EPRI AP-5480

European Commission (2015) Ecological flows in the implementation of the Water Framework Directive. Guidance Document No. 31. European Commission Technical Report 2015-086. https://doi.org/10.2779/775712

Fantin-Cruz I, Pedrollo O, Girard P, Zeilhofer P, Hamilton SK (2016) Changes in river water quality caused by a diversion hydropower dam bordering the Pantanal floodplain. Hydrobiologia 768(1):223–238. https://doi.org/10.1007/s10750-015-2550-4

Ferguson JW, Absolon RF, Carlson TJ, Sandford BP (2006) Evidence of delayed mortality on juvenile pacific salmon passing through turbines at Columbia river dams. Trans Am Fish Soc 135(1):139–150. https://doi.org/10.1577/t05-080.1

Franssen NR, Tobler M (2013) Upstream effects of a reservoir on fish assemblages 45 years following impoundment. J Fish Biol 82(5):1659–1670

Freyhof J, Gessner M, Grossart HP, Hilt S, Jähnig S, Köhler J, Mehner T, Pusch M, Venohr M, Wolter C (2019) Strengths and weaknesses of the Water Framework Directive (WFD) IGB Policy Brief. https://doi.org/10.4126/FRL01-006416917

Gandini CV, Sampaio FAC, Pompeu PS (2014) Hydropeaking effects of on the diet of a Neotropical fish community. Neotrop Ichthyol 12(4):795–802. https://doi.org/10.1590/1982-0224-20130151

Gardner CJ, Rees-Jones J, Morris G, Bryant PG, Lucas MC (2016) The influence of sluice gate operation on the migratory behaviour of Atlantic salmon Salmo salar (L.) smolts. J Ecohydraul 1(1–2):90–101. https://doi.org/10.1080/24705357.2016.1252251

Gasparatos A, Doll CNH, Esteban M, Ahmed A (2017) Renewable energy and biodiversity: implications for transitioning to a green economy. Renew Sustain Energy Rev 70:161–184. https://doi.org/10.1016/j.rser.2016.08.030

Gehrke PC, Gilligan DM, Barwick M (2002) Changes in fish communities of the Shoalhaven River 20 years after construction of Tallowa Dam Australia. River Res Appl 18(3):265–286

Geist J (2021) Editorial: Green or red: challenges for fish and freshwater biodiversity conservation related to hydropower. Aquat Conserv Mar Freshwat Ecosyst 31(7):1551–1558. https://doi.org/10.1002/aqc.3597

Gloss SP, Wahl JR (1983) Mortality of Juvenile salmonids passing through Ossberger crossflow turbines at small-scale hydroelectric sites. Trans Am Fish Soc 112(2A):194–200. https://doi.org/10.1577/1548-8659(1983)112%3c194:mojspt%3e2.0.co;2

Gosset C, Travade F, Durif C, Rives J, Elie P (2005) Tests of two types of bypass for downstream migration of eels at a small hydroelectric power plant. River Res Appl 21(10):1095–1105. https://doi.org/10.1002/rra.871

Gowans ARD, Armstrong JD, Priede IG, Mckelvey S (2003) Movements of Atlantic salmon migrating upstream through a fish-pass complex in Scotland. Ecol Freshw Fish 12(3):177–189. https://doi.org/10.1034/j.1600-0633.2003.00018.x

Green RH (1979) Sampling design and statistical methods for environmental biologists. John Wiley & Sons

Habit E, Belk MC, Parra O (2007) Response of the riverine fish community to the construction and operation of a diversion hydropower plant in central Chile. Aquat Conserv Mar Freshwat Ecosyst 17(1):37–49. https://doi.org/10.1002/aqc.774

Harrison PM, Martins EG, Algera DA, Rytwinski T, Mossop B, Leake AJ, Power M, Cooke SJ (2019) Turbine entrainment and passage of potadromous fish through hydropower dams: developing conceptual frameworks and metrics for moving beyond turbine passage mortality. Fish Fish 20(3):403–418. https://doi.org/10.1111/faf.12349

Havn TB, Sæther SA, Thorstad EB, Teichert MAK, Heermann L, Diserud OH, Borcherding J, Tambets M, Økland F (2017) Downstream migration of Atlantic salmon smolts past a low head hydropower station equippped with Archimedes screw and Francis turbines. Ecol Eng 105:262–275. https://doi.org/10.1016/j.ecoleng.2017.04.043

Hayes DS, Moreira M, Boavida I, Haslauer M, Unfer G, Zeiringer B, Greimel F, Auer S, Ferreira T, Schmutz S (2019) Life stage-specific hydropeaking flow rules. Sustainability (switz#erland) 11(6):1547. https://doi.org/10.3390/su11061547

Hecht JS, Lacombe G, Arias ME, Dang TD, Piman T (2019) Hydropower dams of the Mekong River basin: a review of their hydrological impacts. J Hydrol 568:285–300. https://doi.org/10.1016/j.jhydrol.2018.10.045

Hedger RD, Sauterleute J, Sundt-Hansen LE, Forseth T, Ugedal O, Diserud OH, Bakken TH (2018) Modelling the effect of hydropeaking-induced stranding mortality on Atlantic salmon population abundance. Ecohydrology 11(5). https://doi.org/10.1002/eco.1960

Heisey PG, Mathur P, Euston ET (1996) Passing fish safely: a closer look at turbine vs. spillway survival. Hydro Rev 35(4)

Hershey H (2021) Updating the consensus on fishway efficiency: a meta-analysis. February, 1–14. https://doi.org/10.1111/faf.12547

Hogan TW, Čada GF, Amaral SV (2014) The status of environmentally enhanced hydropower turbines. Fisheries 39(4):164–172. https://doi.org/10.1080/03632415.2014.897195

Huckstorf V, Lewin WC, Wolter C (2008) Environmental flow methodologies to protect fisheries resources in human-modified large lowland rivers. River Res Appl 24(5):519–527. https://doi.org/10.1002/rra.1131

Jackson DC (1985) The influence of differing flow regimes on the Coosa River tailwater fishery below Jordan Dam. Diss Abst Int Pt B-Sci Eng 6(6):113

Jackson DC, Marmulla G (2001) The influence of dams on river fisheries. FAO Fish Tech Pap 419:1–44

Jepsen N, Pedersen S, Thorstad E (2000) Behavioural interactions between prey (trout smolts) and predators (pike and pikeperch) in an impounded river. Regul Rivers: Res Manage 16(2):189–198. https://doi.org/10.1002/(sici)1099-1646(200003/04)16:2%3c189::aid-rrr570%3e3.3.co;2-e

Jonsson B, Waples RS, Friedland KD (1999) Extinction considerations for diadromous fishes. ICES J Mar Sci 56(4):405–409. https://doi.org/10.1006/jmsc.1999.0483

Jungwirth M, Haidvogl G, Moog O, Muhar S, Schmutz S (2003) Angewandte Fischökologie an Fließgewässern. Facultas-Verlag, 547

Jungwirth M, Schmutz S, Weiss, S (1998) Fish migration and fish bypasses (Vol. 4). Fishing News Books Oxford

Katano O, Nakamura T, Abe S, Yamamoto S, Baba Y (2006) Comparison of fish communities between above-and below-dam sections of small streams; barrier effect to diadromous fishes. J Fish Biol 68(3):767–782

Katopodis C (1992) Introduction to fishway design. In Oceans (Issue January). Freshwater Institute, Central and Arctic Region, Department of Fisheries. http://www.wra.gov.tw/public/attachment/41110254871.pdf

Katopodis C, Williams JG (2012) The development of fish passage research in a historical context. Ecol Eng 48:8–18

Kemp PS, Russon IJ, Vowles AS, Lucas MC (2011) The influence of discharge and temperature on the ability of upstream migrant adult river lamprey (Lampetra fluviatilis) to pass experimental overshot and undershot weirs. River Res Appl 27(4):488–498

Larinier M (2001) Environmental issues, dams and fish migration. FAO Fish Tech Pap 419:45–90

Larinier M, Travade F (1992) La conception des dispositifs de franchissement pour les aloses. Bulletin Français De La Pêche Et De La Pisciculture 326–327:125–133. https://doi.org/10.1051/kmae:1992009

Larinier M, Travade F (2002) Downstream migration: problems and facilities. Bulletin Français De La Pêche Et De La Pisciculture 364:181–207

Lees AC, Peres CA, Fearnside PM, Schneider M, Zuanon JAS (2016) Hydropower and the future of Amazonian biodiversity. Biodivers Conserv 25(3):451–466. https://doi.org/10.1007/s10531-016-1072-3

Lucas MC, Bubb DH, Jang M-H, Ha K, Masters JEG (2009) Availability of and access to critical habitats in regulated rivers: effects of low-head barriers on threatened lampreys. Freshw Biol 54(3):621–634. https://doi.org/10.1111/j.1365-2427.2008.02136.x

Lucas MC, Frear PA (1997) Effects of a flow-gauging weir on the migratory behaviour of adult barbel, a riverine cyprinid. J Fish Biol 50(2):382–396. https://doi.org/10.1006/jfbi.1996.0302

Martinez PJ, Chart TE, Trammell MA, Wullschleger JG, Bergersen EP (1994) Fish species composition before and after construction of a main stem reservoir on the White River Colorado. Environ Biol Fishes 40(3):227–239

Martins EG, Gutowsky LFG, Harrison PM, Patterson DA, Power M, Zhu DZ, Leake A, Cooke SJ (2013) Forebay use and entrainment rates of resident adult fish in a large hydropower reservoir. Aquat Biol 19(3):253–263. https://doi.org/10.3354/ab00536

Marttin F, De Graaf GJ (2002) The effect of a sluice gate and its mode of operation on mortality of drifting fish larvae in Bangladesh. Fish Manage Ecol 9(2):123–125

Mathur D, Heisey PG, Skalski JR, Kenney DR (2000) Salmonid smolt survival relative to turbine efficiency and entrainment depth in hydroelectric power generation 1. JAWRA J Am Water Resourc Assoc 36(4):737–747

Matt P, Pirker O, Schletterer M (2019). Hydropower through time – The significance of Alpine rivers for the energy sector. In: Muhar S, Huhar A, Egger G, Siegrist D (eds) Rivers of the Alps—diversity in nature and culture, pp 248–259

McCarthy TK, Frankiewicz P, Cullen P, Blaszkowski M, O'Connor W, Doherty D (2008) Long-term effects of hydropower installations and associated river regulation on River Shannon eel populations: mitigation and management. Hydrobiologia 609(1):109–124. https://doi.org/10.1007/s10 750-008-9395-z

Moreira M, Hayes DS, Boavida I, Schletterer M, Schmutz S, Pinheiro A (2019) Ecologically-based criteria for hydropeaking mitigation: a review. Sci Total Environ 657:1508–1522. https://doi.org/ 10.1016/j.scitotenv.2018.12.107

Morita K, Yamamoto S (2002) Effects of habitat fragmentation by damming on the persistence of stream-dwelling charr populations. Conserv Biol 16(5):1318–1323. https://doi.org/10.1046/j. 1523-1739.2002.01476.x

Mueller M, Bierschenk AM, Bierschenk BM, Pander J, Geist J (2020a) Effects of multiple stressors on the distribution of fish communities in 203 headwater streams of Rhine, Elbe and Danube. Sci Total Environ 703:134523

Mueller M, Knott J, Pander J, Geist J (2020b) Fischökologisches Monitoring an innovativen Wasserkraftanlagen Abschlussbericht 2020b Band 3: Baiersdorf-Wellerstadt an der Regnitz. Lehrstuhl für Aquatische Systembiologie, Technische Universität München, Wissenschaftszentrum Weihenstephan

Mueller M, Knott J, Pander J, Geist J (2020c) Fischökologisches Monitoring an innovativen Wasserkraftanlagen Abschlussbericht 2020c Band 4: Lindesmühle an der Fränkischen Saale. Lehrstuhl für Aquatische Systembiologie, Technische Universität München, Wissenschaftszentrum Weihenstephan

Mueller M, Knott J, Pander J, Geist J (2020d) Fischökologisches Monitoring an innovativen Wasserkraftanlagen Abschlussbericht 2020d Band 5: Au an der Iller. Lehrstuhl für Aquatische Systembiologie, Technische Universität München, Wissenschaftszentrum Weihenstephan

Mueller M, Knott J, Pander J, Geist J (2020e) Fischökologisches Monitoring an innovativen Wasserkraftanlagen Abschlussbericht 2020e Band 6: Heckerwehr an der Roth. Lehrstuhl für Aquatische Systembiologie, Technische Universität München, Wissenschaftszentrum Weihenstephan

Mueller M, Knott J, Pander J, Geist J (2020f) Fischökologisches Monitoring an innovativen Wasserkraftanlagen Abschlussbericht 2020f Band 7: Eixendorf an der Schwarzach. Lehrstuhl für Aquatische Systembiologie, Technische Universität München, Wissenschaftszentrum Weihenstephan

Mueller M, Knott J, Pander J, Geist J (2020g) Fischökologisches Monitoring an innovativen Wasserkraftanlagen Abschlussbericht 2020g Band 8: Baierbrunn an der Isar. Lehrstuhl für Aquatische Systembiologie, Technische Universität München, Wissenschaftszentrum Weihenstephan

Mueller M, Knott J, Pander J, Geist J (2020h) Fischökologisches Monitoring an innovativen Wasserkraftanlagen Abschlussbericht 2020h Band 9: Höllthal an der Alz. Lehrstuhl für Aquatische Systembiologie, Technische Universität München, Wissenschaftszentrum Weihenstephan

Mueller M, Pander J, Geist J (2011) The effects of weirs on structural stream habitat and biological communities. J Appl Ecol 48(6):1450–1461. https://doi.org/10.1111/j.1365-2664.2011.02035.x

Mueller M, Pander J, Geist J (2017) Evaluation of external fish injury caused by hydropower plants based on a novel field-based protocol. Fish Manage Ecol 24(3):240–255. https://doi.org/10.1111/fme.12229

Mueller M, Sternecker K, Milz S, Geist J (2020i) Assessing turbine passage effects on internal fish injury and delayed mortality using X-ray imaging. PeerJ 8:e9977

Muir WD, Marsh DM, Sandford BP, Smith SG, Williams JG (2006) Post-hydropower system delayed mortality of transported snake river stream-type chinook salmon: unraveling the mystery. Trans Am Fish Soc 135(6):1523–1534. https://doi.org/10.1577/t06-049.1

Muir WD, Smith SG, Williams JG, Sandford BP (2001) Survival of Juvenile Salmonids passing through bypass systems, turbines, and spillways with and without flow deflectors at snake river dams. North Am J Fish Manage 21(1):135–146. https://doi.org/10.1577/1548-8675(2001)021%3c0135:sojspt%3e2.0.co;2

Nijboer RC, Johnson RK, Verdonschot PFM, Sommerhäuser M, Buffagni A (2004) Establishing reference conditions for European streams. Hydrobiologia 516(1–3):91–105. https://doi.org/10.1023/B:HYDR.0000025260.30930.f4

Noonan MJ, Grant JWA, Jackson CD (2012) A quantitative assessment of fish passage efficiency. Fish Fish 13(4):450–464. https://doi.org/10.1111/j.1467-2979.2011.00445.x

Nyqvist D, Elghagen J, Heiss M, Calles O (2018) An angled rack with a bypass and a nature-like fishway pass Atlantic salmon smolts downstream at a hydropower dam. Mar Freshw Res 69(12):1894–1904. https://doi.org/10.1071/MF18065

Odeh M (1999) A summary of environmentally friendly turbine design concepts. U.S. Department of Energy, Idaho Operations Office, Idaho Falls, ID, Leetown Science Center, Report 99-065/TF

Økland F, Havn TB, Thorstad EB, Heermann L, Sæther SA, Tambets M, Teichert MAK, Borcherding J (2019) Mortality of downstream migrating European eel at power stations can be low when turbine mortality is eliminated by protection measures and safe bypass routes are available. Int Rev Hydrobiol 104(3–4):68–79. https://doi.org/10.1002/iroh.201801975

Ovidio M, Dierckx A, Bunel S, Grandry L, Spronck C, Benitez JP (2017) Poor performance of a retrofitted downstream bypass revealed by the analysis of approaching behaviour in combination with a trapping system. River Res Appl 33(1):27–36. https://doi.org/10.1002/rra.3062

Pelicice FM, Agostinho AA (2009) Fish fauna destruction after the introduction of a non-native predator (Cichla kelberi) in a Neotropical reservoir. Biol Invasions 11(8):1789–1801

Pelicice FM, Pompeu PS, Agostinho AA (2015) Large reservoirs as ecological barriers to downstream movements of Neotropical migratory fish. Fish Fish 16(4):697–715. https://doi.org/10.1111/faf.12089

Penczak T, Głowacki Ł, Galicka W, Koszaliński H (1998) A long-term study (1985–1995) of fish populations in the impounded Warta River Poland. Hydrobiologia 368(1–3):157–173. https://doi.org/10.1023/A:1003246115666

Person É (2013) Impact of hydropeaking on fish and their habitat. In Communications du Laboratoire de Constructions Hydrauliques—55 (vol 5812, Issue 2013). EPFL-LCH. https://doi.org/10.5075/epfl-thesis-5812

Pflugrath BD, Boys CA, Cathers B (2019) Over or under? Autonomous sensor fish reveals why overshot weirs may be safer than undershot weirs for fish passage. Ecol Eng 132:41–48. https://doi.org/10.1016/j.ecoleng.2019.03.010

Piorkowski RJ (1995). Ecological effects of spawning salmon on several southcentral Alaskan streams

Piper AT, Rosewarne PJ, Wright RM, Kemp PS (2018) The impact of an Archimedes screw hydropower turbine on fish migration in a lowland river. Ecol Eng 118:31–42. https://doi.org/10.1016/j.ecoleng.2018.04.009

Pracheil BM, DeRolph CR, Schramm MP, Bevelhimer MS (2016) A fish-eye view of riverine hydropower systems: the current understanding of the biological response to turbine passage. Rev Fish Biol Fisheries 26(2):153–167. https://doi.org/10.1007/s11160-015-9416-8

Pulg U, Schnell J (2008) Untersuchungen zur Effektivität alternativer Triebwerkstechniken und Schutzkonzepte für abwandernde Fische beim Betrieb von Kleinwasserkraftanlagen. In Landesfischereiverband Bayern

Quaranta E, Wolter C (2021) Sustainability assessment of hydropower water wheels with downstream migrating fish and blade strike modelling. Sustain Energy Technol Assess 43:100943

Reed BC, Kelso WE, Rutherford DA (1992) Growth, fecundity, and mortality of Paddlefish in Louisiana. Trans Am Fish Soc 121(3):378–384. https://doi.org/10.1577/1548-8659(1992)121%3c0378:gfamop%3e2.3.co;2

Reid AJ, Carlson AK, Creed IF, Eliason EJ, Gell PA, Johnson PTJ, Kidd KA, Maccormack TJ, Olden JD, Ormerod SJ, Smol JP, Taylor WW, Tockner K, Vermaire JC, Dudgeon D, Cooke SJ (2019) Emerging threats and persistent conservation challenges for freshwater biodiversity. 94:849–873. https://doi.org/10.1111/brv.12480

Reuter M, Kohout C (2014) Praxishandbuch für den umweltbewussten Einsatz von Turbinentechnologien im Bereich der Kleinstwasserkraft. Institut für Wasserwirtschaft, Siedlungswasserbau und Ökologie GmbH, Hydrolabor Schleusingen, Schleusingen

Richmond MC, Serkowski JA, Ebner LL, Sick M, Brown RS, Carlson TJ (2014) Quantifying barotrauma risk to juvenile fish during hydro-turbine passage. Fish Res 154:152–164. https://doi.org/10.1016/j.fishres.2014.01.007

Roscoe DW, Hinch SG, Cooke SJ, Patterson DA (2011) Fishway passage and post-passage mortality of up-river migrating sockeye salmon in the Seton River British Columbia. River Res Appl 27(6):693–705. https://doi.org/10.1002/rra.1384

Ruggles CP, Murray DG (1983) A Review of Fish Response to spillways. Can Tech Rep Fisher Aquat Sci 1172:29p

Santos JM, Branco P, Katopodis C, Ferreira T, Pinheiro A (2014) Retrofitting pool-and-weir fishways to improve passage performance of benthic fishes: effect of boulder density and fishway discharge. Ecol Eng 73:335–344. https://doi.org/10.1016/j.ecoleng.2014.09.065

Santos JM, Ferreira MT, Pinheiro AN, Bochechas JH (2006) Effects of small hydropower plants on fish assemblages in medium-sized streams in central and northern Portugal. Aquat Conserv Mar Freshwat Ecosyst 16(4):373–388. https://doi.org/10.1002/aqc.735

Sá-Oliveira JC, Hawe JE, Isaac-Nahum VJ, Peres CA (2015) Upstream and downstream responses of fish assemblages to an eastern Amazonian hydroelectric dam. Freshw Biol 60(10):2037–2050. https://doi.org/10.1111/fwb.12628

Schiemer F, Keckeis H, Winkler G, Flore L (2001) Large rivers: the relevance of ecotonal structure and hydrological properties for the fish fauna. River Syst 12(2–4):487–508. https://doi.org/10.1127/lr/12/2001/487

Schilt CR (2007) Developing fish passage and protection at hydropower dams. Appl Anim Behav Sci 104(3–4):295–325. https://doi.org/10.1016/j.applanim.2006.09.004

Schletterer M, Reindl R, Thonhauser T (2016) Options for re-establishing river continuity, with an emphasis on the special solution "fish lift": examples from Austria. Revista Eletrônica de Gestão e Tecnologias Ambientais, 109–128

Schmidt MB, Tuhtan JA, Schletterer M (2018) Hydroacoustic and pressure turbulence analysis for the assessment of fish presence and behavior upstream of a vertical trash rack at a run-of-river hydropower plant. Appl Sci (switzerland) 8(10):1723. https://doi.org/10.3390/app8101723

Schmutz S, Bakken TH, Friedrich T, Greimel F, Harby A, Jungwirth M, Melcher A, Unfer G, Zeiringer B (2015) Response of fish communities to hydrological and morphological alterations in hydropeaking rivers of Austria. River Res Appl 31(8):919–930. https://doi.org/10.1002/rra.2795

Schmutz S, Sendzimir J (2018) Riverine ecosystem management: science for governing towards a sustainable future. Springer Nature

Shen Y, Diplas P (2010) Modeling unsteady flow characteristics of hydropeaking operations and their implications on fish habitat. J Hydraul Eng 136(12):1053–1066. https://doi.org/10.1061/(ASCE)HY.1943-7900.0000112

Silva S, Barca S, Vieira-Lanero R, Cobo F (2019) Upstream migration of the anadromous sea lamprey (Petromyzon marinus Linnaeus, 1758) in a highly impounded river: impact of low-head obstacles and fisheries. Aquat Conserv Mar Freshwat Ecosyst 29(3):389–396. https://doi.org/10.1002/aqc.3059

Smith EP (2014) BACI design. Wiley StatsRef: Statistics Reference Online

Stansell RJ, Gibbons KM, Nagy WT (2010) Evaluation of pinniped predation on adult salmonids and other fish in the Bonneville Dam tailrace, 2008–2010

Stendera S, Adrian R, Bonada N, Cañedo-Argüelles M, Hugueny B, Januschke K, Pletterbauer F, Hering D (2012) Drivers and stressors of freshwater biodiversity patterns across different ecosystems and scales: a review. Hydrobiologia 696(1):1–28. https://doi.org/10.1007/s10750-012-1183-0

Stephenson JR, Gingerich AJ, Brown RS, Pflugrath BD, Deng Z, Carlson TJ, Langeslay MJ, Ahmann ML, Johnson RL, Seaburg AG (2010) Assessing barotrauma in neutrally and negatively buoyant juvenile salmonids exposed to simulated hydro-turbine passage using a mobile aquatic barotrauma laboratory. Fish Res 106(3):271–278

Stich DS, Zydlewski GB, Kocik JF, Zydlewski JD (2015) Linking behavior, physiology, and survival of Atlantic salmon Smolts during estuary migration. Mar Coast Fisher 7(1):68–86. https://doi.org/10.1080/19425120.2015.1007185

Taylor RE, Kynard B (1985) Mortality of Juvenile American Shad and Blueback Herring passed through a low-head Kaplan hydroelectric turbine. Trans Am Fish Soc 114(3):430–435. https://doi.org/10.1577/1548-8659(1985)114%3c430:mojasa%3e2.0.co;2

Thorne RE, Johnson GE (1993) A review of hydroacoustic studies for estimation of salmonid downriver migration past hydroelectric facilities on the Columbia and Snake Rivers in the 1980s. Rev Fish Sci 1(1):27–56

Thornton KW, Kimmel BL, Payne FE (1990) Reservoir limnology: ecological perspectives. John Wiley & Sons

Thorstad EB, Økland F, Aarestrup K, Heggberget TG (2008) Factors affecting the within-river spawning migration of Atlantic salmon, with emphasis on human impacts. Rev Fish Biol Fisher 18(4):345–371. https://doi.org/10.1007/s11160-007-9076-4

Tiffan KF, Hatten JR, Trachtenbarg DA (2016) Assessing juvenile salmon rearing habitat and associated predation risk in a lower snake river reservoir. 1038:1030–1038. https://doi.org/10.1002/rra

Travade F, Larinier M (2002) Fish locks and fish lifts. Bulletin Français de La Pêche et de La Pisciculture, 364 supplément, 102–118. https://doi.org/10.1051/kmae/2002096

Tuhtan JA, Noack M, Wieprecht S (2012) Estimating stranding risk due to hydropeaking for juvenile European grayling considering river morphology. KSCE J Civ Eng 16(2):197–206. https://doi.org/10.1007/s12205-012-0002-5

Tundisi JG, Straškraba M (1999) Theoretical reservoir ecology and its applications. International Institute of Ecology Ann Arbor

Turnpenny AWH, Clough S, Hanson KP, Ramsay R, McEwan D (2000) Risk assessment for fish passage through small, low-head turbines. Fawley Aquatic Research Laboratories, London, Report ETSU H/06/00054/REP

USFWS (U.S. Fish and Wildlife Service) (2019) Fish passage engineering design criteria. USFWS, Northeast Region R5, Hadley, Massachusetts, 5, 224

Vannote RL, Minshall GW, Cummins KW, Sedell JR, Cushing CE (1980) The river continuum concept. Can J Fish Aquat Sci 37(1):130–137. https://doi.org/10.1139/f80-017

Weibel U (1991) Neue Ergebnisse zur Fischfauna des nördlichen Oberrheins ermittelt im Rechengut von Kraftwerken. Fischökologie 5:43–68

Williams JG, Smith SG, Muir WD (2001) Survival estimates for downstream migrant yearling Juvenile Salmonids through the Snake and Columbia rivers hydropower system, 1966–1980 and 1993–1999. North Am J Fish Manage 21(2):310–317. https://doi.org/10.1577/1548-8675(2001)021%3c0310:sefdmy%3e2.0.co;2

Winter HV, Jansen HM, Bruijs MCM (2006) Assessing the impact of hydropower and fisheries on downstream migrating silver eel, Anguilla anguilla, by telemetry in the River Meuse. Ecol Freshw Fish 15(2):221–228. https://doi.org/10.1111/j.1600-0633.2006.00154.x

Wolter C, Bernotat D, Gessner J, Brüning A, Lackemann J, Radinger J (2020) Fachplanerische Bewertung der Mortalität von Fischen an Wasserkraftanlagen. Bundesamt für Naturschutz

Wood PJ, Armitage PD (1997) Biological effects of fine sediment in the lotic environment. 21(2):203–217

Yang N, Li Y, Zhang W, Lin L, Qian B, Wang L, Niu L, Zhang H (2020). Cascade dam impoundments restrain the trophic transfer efficiencies in benthic microbial food web. Water Res 170:115351

Young PS, Cech JJ, Thompson LC (2011) Hydropower-related pulsed-flow impacts on stream fishes: a brief review, conceptual model, knowledge gaps, and research needs. Rev Fish Biol Fisher 21(4):713–731. https://doi.org/10.1007/s11160-011-9211-0

Ziv G, Baran E, Nam S, Rodríguez-Iturbe I, Levin SA (2012) Trading-off fish biodiversity, food security, and hydropower in the Mekong River Basin. Proc Natl Acad Sci USA 109(15):5609–5614. https://doi.org/10.1073/pnas.1201423109

The Attractiveness of Fishways and Bypass Facilities

5

Armin Peter, Nils Schoelzel, Lisa Wilmsmeier, Ismail Albayrak,
Francisco Javier Bravo-Córdoba, Ana García-Vega,
Juan Francisco Fuentes-Pérez, Jorge Valbuena-Castro,
Omar Carazo-Cea, Carlos Escudero-Ortega,
Francisco Javier Sanz-Ronda, Damien Calluaud, Gérard Pineau,
and Laurent David

A. Peter · N. Schoelzel · L. Wilmsmeier
FishConsulting GmbH, Olten, Switzerland
e-mail: apeter@fishconsulting.ch

N. Schoelzel
e-mail: nils.schoelzel@fishconsulting.ch

L. Wilmsmeier
e-mail: lisa.wilmsmeier@fishconsulting.ch

I. Albayrak (✉)
Laboratory of Hydraulics, Hydrology and Glaciology, ETH Zurich, Zurich, Switzerland
e-mail: albayrak@vaw.baug.ethz.ch

F. J. Bravo-Córdoba · A. García-Vega · J. F. Fuentes-Pérez · J. Valbuena-Castro · O. Carazo-Cea ·
C. Escudero-Ortega · F. J. Sanz-Ronda
ITAGRA – University of Valladolid, Palencia, Spain
e-mail: francisco.bravo@iaf.uva.es

A. García-Vega
e-mail: ana.garcia.vega@iaf.uva.es

J. F. Fuentes-Pérez
e-mail: jfuentes@iaf.uva.es

J. Valbuena-Castro
e-mail: jvalbuena@uva.es

F. J. Sanz-Ronda
e-mail: jsanz@uva.es

P. Rutschmann et al. (eds.), *Novel Developments for Sustainable Hydropower*,
https://doi.org/10.1007/978-3-030-99138-8_5

5.1   Passability: Fish Swimming Behaviour and Fishpass Preferences (e.g. Schiffmühle)

5.1.1   Introduction

Migration and movements of fishes can be observed on a daily, seasonal or annual basis. Individuals move and migrate over short or long distances. Movement is considered to be a change of location or habitat within a river, and is mainly to seek basic resources (food, shelter) or as a reaction to avoid predators. Dingle (2014) emphasizes that most movements take place within a defined area or home range. Foraging (movement in search of resources) or commuting (movement in search of resources, typically daily) are typical examples (Dingle 2014). The size of the home range depends on an individual's swimming ability and behaviour. In general, movement is a repetitive behaviour in time and space. Migration, however, is different from movement. Migration is not carried out to seek available resources (food, habitat), but is a distinct and specialized behaviour involving leaving one habitat and settling in another outside the home range (Dingle 2014). Cyprinid fishes (Lucas and Baras 2001) and salmonids change habitats often during their life cycle, and both show homing behaviour, i.e. the behaviour of spawners returning to the streams or spawning areas in which they spent their early life stages and which is therefore considered to be a suitable habitat for reproduction (Wootton 1990). To fulfil their natural life cycle, fishes depend on an intact migration corridor that is not artificially fragmented.

Considering the need for movement and migration, it is very important for hydropower to be produced in a way that allows fish to use the various habitats and to switch between them in a river system. River fragmentation by dams and its effects on fishes is a worldwide challenge.

New legislation in EU and Switzerland therefore requires fish migration facilities at hydropower plants (HPP) to be built according to the latest standards, focusing on target species while also allowing non-target species to fully ascend the devices built for upstream migration (fish ladders, nature-like bypass systems) or descend them (physical screens and angled bar racks with bypasses).

D. Calluaud · G. Pineau · L. David
Institut Pprime, Pole ecohydraulique OFB/IMFT/Pprime Poitiers, CNRS University of Poitiers, Poitiers, France
e-mail: damien.calluaud@univ-poitiers.fr

G. Pineau
e-mail: gerard.pineau@univ-poitiers.fr

L. David
e-mail: laurent.david@univ-poitiers.fr

Success monitoring must also be carried out in order to verify the fish passage efficiency of upstream and downstream fishpasses. Evaluating a fishway's efficiency after construction is crucial ensuring that the structure serves its purpose and for making any necessary adjustments (Noonan et al. 2012). Therefore, this study deals with the performance evaluation of upstream and downstream fishpasses installed at the case study HPP Schiffmühle in Switzerland by conducting a fish monitoring.

5.1.2 Case Study Hydropower Plant Schiffmühle, River Limmat, Switzerland

There are two run-of-river HPPs Schiffmühle on the River Limmat, which is the outflow of Lake Zurich: the main powerhouse and the residual flow HPPs. The residual flow HPP served as a case study in the FIThydro project (https://www.fithydro.eu/schiffmuhle/) and all the descriptions below refer to this plant. The design discharge of the HPP is 14 m³/s, and the head is 3.17 m. Its electricity generation capacity is 0.5 MW (bevel gear bulb turbine). The 75 m long upstream migration facility for upstream migrating fishes is a combination of a nature-like fishpass (NL) and a vertical-slot (VS) fishpass (Figs. 5.1 and 5.2). The entrance into the NL pass is 36 m downstream of the turbine flow outlet (entrance angle about 40 degrees). The entrance into the VL pass, which has a 6.3% slope, is located 2 m downstream of the turbine flow, with an entrance angle of 90 degrees. Total discharge of the pass is 500 l/s. After 10 pools in the VS pass and 12 pools in the NL pass, the two passes merge and an additional 12 pools follow in the upstream direction until the exit.

The technical device for downstream migration is a Horizontal Bar Rack-Bypass System (HBR-BS) (Figs. 5.1 and 5.2, also see Sect. 7.2). The rack is positioned almost parallel to the main flow, with a lateral intake. The rack is 14.6 m long and 1.82 m high, and the clear bar spacing is 20 mm. The bars have rectangular profiles. The average velocity in front of the screen is 0.5 m/s at the design discharge. It is intended to guide fish into a bypass with two openings in a vertical chamber at different water depths (bottom and water surface). There is a 25 cm diameter pipe bypass to lead the fish to the downstream part of the HPP. The discharge of the bypass pipe is 170 l/s.

This study answers the following research questions:

- What is the passage efficiency of fishpasses for upstream migration fishes and how much time does a successful passage take?
- Do fishes have a special preference for one of the two entrances?
- Are all fish species and all fish sizes able to negotiate the fish passes successfully, and what are their migration seasons?
- Are downstream migrating individuals able to find the entrance into the bypass?

Fig. 5.1 HPP Schiffmühle, River Limmat. VS: vertical-slot pass, NL: nature-like pass, EX: exit of fishpass, CF: counting facility of the fishpass, HS: horizontal bar rack, ENBY: entrance bypass, EXBY: exit bypass

5.1.3 Methods

PIT-tagging was used to detect the migration of fishes. Two antennae were installed at each entrance, one of them as close as possible to the entrance, and a second in the fourth or fifth pool of the ladder. One additional antenna was installed in the upper part of the fish ladder where fish leave the fishpass. In order to detect downstream migration, one antenna was placed in the bypass pipe close to the entrance.

Individuals were tagged with HDX (half-duplex) Pit-tags (Oregon RFID, Texas Instruments ISO 11784/11785) and measured (total length). 23 mm (individuals > 150 mm) and 12 mm tags were used. A total of 3087 individuals were tagged between September 2017 and September 2019. Tagged fish were allowed to recover and then released 210 m downstream of the nature-like fishpass entrance on the left riverbank. Data treatment was carried out with R (R development core team 2013).

Attraction efficiency was defined as the proportion of individuals tagged that were subsequently detected at the first antenna at the fishpass entrance. Entrance efficiency

Fig. 5.2 **a** PIT-tagging antenna in the nature-like fishpass, **b** antenna in the vertical-slot pass, **c** turbine inlet with horizontal bar rack and **d** close-up photo of bypass system

was the number of fish detected at the second antenna divided by the number of fish registered at the entrance antenna.

Fish passage efficiency was defined by dividing the number of fish exiting the fishway by the number of fish detected at the second antenna at the pass entrance (Bunt et al. 2012). Passage time was calculated as the time from the last detection at the entrance antennae to successful exit.

5.1.4 Passage Efficiency

A total of 13 fish species were tagged within the group of the 2890 individuals caught in the pass counting basin. The rheophilic barbel (*Barbus barbus*), chub (*Squalius cephalus*) and spirlin (*Alburnoides bipunctatus*) representative of schooling fishes, were the most abundant individuals tagged. Other tagged fish species were brown trout, perch (*Perca fluviatilis*), bullhead, gudgeon, dace, bleak (*Alburnus alburnus*), nase (*Chondrostoma nasus*), roach, rudd and pumpkinseed.

Most of the tagged fish were between 90 and 260 mm long. 67.3% of the tagged individuals were redetected after tagging (attraction efficiency), 95.5% of them entered the fishpasses. Other studies have documented smaller attraction values (32.9% by Benitez et al. 2018), or approximately 30% for all entrances at the HPP Rheinfelden in the River Rhine (Peter et al. 2016). In contrast, Noonan et al. (2012) published average attraction efficiencies of 65.1%; however, most of these studies were carried out on salmonids, which have a high motivation to migrate. A meta-analysis by Bunt et al. (2012) also documented average attraction efficiencies of 66% (different fish species and types of migration facilities). The obtained values in the Limmat can therefore be classified in the mid-range but clearly higher than in the study in the Rhine. Fish used different entrances into the fishpasses. Bleak showed no preference for one of the entrances, but dace did show a clear preference. The attraction efficiency for dace for the VS fishpass was only 9.8%: conversely, 79.3% preferred the NL fishpass. The same clear preference was also observed for roach (*Rutilus rutilus*, 69.1%). Chub and perch likewise preferred the nature-like fishpass, however distinctly less. Species that preferred the VS fishpass were barbel and, to a lesser extent, spirlin.

The entrance efficiency over all species was 98.5% in the VS fishpass and 87.9% in the NL fishpass, indicating a high efficiency in this study. The passage efficiencies were 78.5% for the VS and 73.6% for the NL fishpasses, respectively and the passage efficiency for the whole fishpass was 81.4% or all species. Such high efficiencies are comparable to the values published by Benitez et al. (2018) with 86.3%, but clearly higher than those published by Noonan et al. (2012, value 41.7%) and Bunt et. al. (2012) (values 45–70%). Finally, the passage efficiencies at the Schiffmühle HPP have to be assessed as good.

5.1.5 Migration Time

Four main periods were observed for the time of migration, coinciding with the time the tagged fish were released. Many ascents were observed after each release of fishes. The four periods were October 2017, May–June 2018, September-mid November 2018, and July–November 2019. The effect of the release date on the date of registration by the antennae was obvious. The tagged and released fish continued their migration without interruption. However, between December and April the antennae registered almost no fishes. The migration activity in winter is thus very low.

The values for the median and the minimum time needed for the passage are very meaningful. For most of the fish species, 50% of all individuals needed less than 60 min for the passage. Spirlin generally took well over 60 min. However, the passage time of the fastest spirlin was only 15 min (Table 5.1).

The passage times in the NL fishpass were somewhat longer than in the VS fishpass. The median passage time increased by 25 min (not including spirlin). The spirlin had a clearly longer passage time. A chub had the fastest passage, with 6.4 min to reach the exit of the fishpass (Table 5.2).

Table 5.1 Duration of passage (in minutes) from the last registration at the lowest antenna in the vertical slot pass to the upper antenna at the exit of the fishpass

Species	Average	Median	Minimum	Maximum	N
eel	76.1	76.1	76.1	76.1	1
chub	77.5	38.0	15.0	927.4	53
brown trout	29.4	29.4	29.4	29.4	1
barbel	217.9	62.6	12.2	28,809.0	434
perch	71.4	59.8	19.6	217.9	17
dace	33.2	29.0	14.2	67.0	6
bleak	70.4	42.3	18.0	580.8	24
roach	61.5	42.4	16.0	429.5	23
rudd	52.8	52.8	52.8	52.8	1
spirlin	1549.6	136.7	15.0	53,952.9	179
pumpkinseed	759.9	759.9	106.4	1413.4	2

Table 5.2 Duration of passage (in minutes) from the last registration at the lowest antenna in the nature-like fish ladder to the upper antenna at the exit of the fishpass

Species	Average	Median	Minimum	Maximum	N
chub	191.1	56.8	6.4	7844.3	182
barbel	252.5	97.4	24.5	8406.1	196
perch	225.9	88.7	31.0	926.6	33
dace	56.0	47.8	22.4	211.3	47
bleak	109.1	62.9	22.2	764.0	27
nase	290.0	290.0	96.3	483.6	2
roach	117.6	69.2	27.6	1408.3	148
spirlin	1225.8	639.5	36.4	21,518.8	77
pumpkinseed	88.4	88.4	88.4	88.4	1

In general, the time needed for the upstream passage is short and the observed passage time of about one hour for more than 50% of the individuals cannot be regarded as an impairment of migration.

5.1.6 Migration Distance

769 fish (25% of the tagged individuals at the HPP Schiffmühle) were detected at the fish-pass facility of the HPP Aue, which is situated 7.03 km upstream of the HPP Schiffmühle.

All of them had to successfully pass the HPP Kappelerhof, which is 2.17 km upstream. Mainly dace (73.5%), roach (62.9%) and nase (60%) tagged at HPP Schiffmühle migrated as far as HPP Aue. However, 37.1% of the bleak also migrated the 7.03 km distance.

5.1.7 Downstream Migration

Over the whole observation period a total of 445 tagged fish descended downstream. Only two fish used the bypass for downstream migration. Other migration corridors were the NL fishpass ($N = 56$ individuals), the VS fishpass ($N = 122$ individuals) and unknown corridors ($N = 265$ individuals). The unknown corridors include the downstream migration over the spillway, through the turbine or through the main HPP at Schiffmühle (Fig. 5.1).

Most of the individuals used unknown migration corridors. The bypass was often clogged with wood affecting the bypass flow and discharge. Furthermore, the attraction flow to the bypass seems inefficient and a recirculation zone possibly affects fish searching and finding the bypass entrance. Moreover, the acceleration at the beginning of the pipe was far too high. The bypass at HPP Schiffmühle is therefore far from fully functional. Fishes migrating downstream used alternative corridors.

5.1.8 Conclusions

Knowledge of the passage efficiency at each single dam is very important for assessing the cumulative impacts of 10 consecutive dams in the River Limmat. Low passage efficiency can have a detrimental impact on populations. PIT-tagging studies are a very useful tool to assess upstream migration facilities. Attraction efficiencies, passage efficiencies and passage time are important parameters for evaluating success. Our results demonstrate that Both vertical slot and nature-like fishpasses at HPP Schiffmühle function well for upstream moving fish with high attractiveness, entrance and passage efficiencies. Species-specific preferences were observed for the available entrances. Having more than one entrance into a fish ladder can therefore be an advantage in rivers with a broad range of species. Tagged fish continued their migration without interruption, proving that the attraction of the fishpass was good. A considerable number of individuals migrated further upstream in the River Limmat over a distance of 5.1 km. An important parameter for assessing swimming performance in the fishpass is the time needed for the passage. The time taken for the ascent was generally low.

The bypass for downstream migration was barely used at all, and serious problems were detected in terms of clogging of the bypass entrance, unfavourable flow conditions around the bypass and high acceleration at the entrance into the pipe. This result indicates that design, location and operation of a bypass system are of prime importance for a successful implementation and high fish guidance efficiency of HBR-BS. Therefore, bypass system needs optimization.

Overall, the present findings have a wide range of applications for other similarly sized HPPs and will serve as a basis for an optimal design of fishpasses and HBR-BS for various fish species.

5.1.9 Acknowledgements

We thank the Limmatkraftwerke AG (Andreas Doessegger, Peter Rotenfluh, Christoph Froelich and his team), the fishing club (Pachtvereinigung Stausee Wettingen) with Reto Wittwer and his team for the support for the fish tagging, and Tabea Kropf (Sektion Jagd und Fischerei des Kantons Aargau).

5.2 Assessing the Effectiveness of Migration Facilities in Guma Iberian Testcase

5.2.1 Introduction

The FIThydro testcase Guma is situated in the Duero River, northwest part of Spain (Fig. 5.3). The Duero River basin presents a high degree of fragmentation with more than 140 small hydropower plants (HPP) (below 10 MW in EU) and 23 large HPP in the Spanish part of the basin, as well as nearly 5200 other obstacles for irrigation, domestic and industrial water supply (www.chduero.es). The presence of these obstacles causes

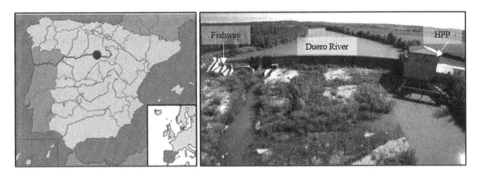

Fig. 5.3 Location of Guma testcase and associated facilities

a high disruption of the fluvial longitudinal connectivity, which hinders or even pre-
vents fish migration, among other associated environmental problems related to discharge
abstraction and sediment management (Branco et al. 2017; Nilsson et al. 2005).

The hydrology of the Duero River in the river section affected by Guma HPP is characterized by low flows in summer (exacerbated by upstream water diversions for irrigation)
and medium to high flows during winter and early spring, associated with the rainy
season and snow-melting episodes. The river reach is dominated by small and medium-
size potamodromous rheophilic endemic cyprinids, such as Iberian barbel (*Luciobarbus
bocagei*), northern straight-mouth nase (*Pseudochondrostoma duriense*), northern Iberian
chub (*Squalius carolitertii*), and Pyrenean gudgeon (*Gobio lozanoi*). Their populations
are currently suffering an important decrease (specially nase, categorized as endangered
(IUCN, 2018)), whereas alien species are increasing their presence (mainly the bleak
(*Alburnus alburnus*)). River discharge during cyprinid upstream migration (April–July)
is between 15 and 30 m^3/s. The river reach belongs to the Epipotamon zone (Illies and
Botoseanu 1963) with an average altitude of around 810 m a.s.l. and corresponds to C6
category (Rosgen and Silvey 1996).

The Guma HPP is operated by Salto de Vadocondes S.A. (SAVASA). It is a run-of-
river HPP with a total height of 8.85 m, an installed capacity of 2.25 MW, and 2 Kaplan
turbines (Fig. 5.3). Due to the run-of-river configuration, there is no legal requirement
for maintaining a minimum environmental flow through the dam. In the right bank of the
dam, there is a pool and weir type fishway composed of 36 cross-walls with submerged
notches and bottom orifices. Through the fishway runs a discharge of 0.25–0.50 m^3/s.
This discharge ensures proper operation of the fishway and it is additionally supported by
a supplementary attraction flow of an additional channel near the fishway entrance (Table
5.3).

The main FIThydro objective in this testcase was to evaluate the effectiveness of the
fishway and the overall performance of the facility for bidirectional fish migration. The
Testcase is representative for the most common HPP configurations in Mediterranean
environments; therefore, its results are of interest to advance in the understanding of
bidirectional fish migration in the Mediterranean area. In addition, this information will
help operators to fulfill the requirements of cost-effective energy production and, at the
same time, meet the environmental requirements and targets under European legislation.

5.2.2 Materials and Methods

To evaluate the effectiveness of migration facilities, native fish species (*Luciobarbus
bocagei, Pseudochondrostoma duriense, Squalius carolitertii*) were PIT tagged. Iberian
barbel individuals represented more than 90% of the sample. Three sets of experiments
were carried out: (1) Fishway evaluation under free conditions (fish released in the river),

Table 5.3 Geometric characteristics and design operating values of the fishway

Fishway type	Submerged Notch with Bottom Orifice
Number of pools	35
Volumetric Power Dissipation	121 ± 10 W/m^3
Slope	8.77%
Pool dimension (length x width)	2.60 m $\times$ 1.60 m
Width of notches	0.30 m
Sill height	0.8 m
Bottom orifice size (height x width)	0.20 m $\times$ 0.20 m
Discharge	0.27 ± 0.01 m^3/s
Water depth	1.32 m
Water drop between pools	0.25 m
Water velocity at notches	1.29 ± 0.07 m/s
Water velocity at orifices	1.94 ± 0.09 m/s

(2) fishway evaluation under confined conditions (fish locked in the fishway), and (3) punctual downstream fish movement analysis in the fishway.

The fish samplings for the first and second experiments were carried out from May to October 2018 and 2019 (at least one per month), collecting fish from different origins (i.e. downstream, upstream, and inside the fishway). In both experiments, fish movements were recorded using a pass-through antenna system, which consisted of four antennas installed in the fishway, covering notch and orifice passage, and connected to a dedicated reader (ORFID® Half Duplex multiplexer reader).

For the first experiments (free conditions), 754 fish were tagged using different release places (411 downstream and 343 upstream, and in both river banks, while in the second set of experiments (confined conditions, n = 153) fish were locked inside the fishway for less than 24 h.

The third set of experiments (downstream movement analysis) was carried out between June and October 2020 (with monthly frequency). They consisted of the capture of all fish inside the fishway, their release downstream outside the fishway, the installation of a close mesh in the middle side turning pool of the fishway, and the counting of fish upstream and downstream of the mesh after 24 h, to identify upstream and downstream movements of fish.

All procedures were carried out following national and community legislation and ethical guidelines about research with animals (Directive 2010/63/UE, and Spanish Act RD 53/2013).

5.2.3 Results

Upstream Migration

For experiments under free conditions, 204 fish from 411 downstream released fish, located and entered the fishway (50% in 2018 and 31% in 2019), of which 129 had success in the fishway ascent (61%). Additionally, 17 fish released in 2018 located and entered the fishway in 2019, which would increase the proportion of location in the subsequent year. Fish spent a median of 9 days (InterQuartile range: 5–19 days) locating the fishway. Both fishway location and passage success were influenced by the origin of the fish. In addition, for the Iberian barbel, fish length significantly affected fishway location (median for those that did not locate the fishway of 123 mm versus 142 mm for successful location); however, fish length did not show significant differences in the ascent success. Fishway location and ascent success were also related to environmental variables, i.e. total river discharge, water level difference between upstream and downstream the dam, and water temperature. There was a period of peak movement mainly during the second half of June and the first days of July, related to changes in the river discharge and water temperature (Fig. 5.4). Regarding the transit time, the global median time for ascending the fishway was 3.4 h (median transit time per meter of height ascended was 26 min/m; 31 pools with water drops of 0.25 m, i.e. a total water head assessed near 8 m). The transit time varied significantly according to the fish origin, which seems to indicate a higher motivation for the ascent of those fish from upstream origin.

For confined experiments, more than 90% of fishpassed the fishway successfully and the median transit time to ascend 2.25 m in height was lower than 23 min (median transit time per meter of height ascended was 10.2 min/m). Fish length had a significant effect on ascent time, with the larger fish the faster to ascend the fishway. Fish in confined experiments were faster than fish in free experiments, presenting speeds similar to those obtained in a vertical slot fishway with the same discharge and under confined conditions (see details in Bravo-Córdoba et al. 2018).

Downstream Migration

The collected data disclosed that the Guma fishway can be used as a downstream migration route. Several fish moved upstream and downstream were identified through the fishway during the tracking period. Regarding fish that were released specifically upstream Guma dam (2018 and 2019), 42% (144/343) were found in the fishway and 64% (92/144) of these fish completed the downstream movement (Fig. 5.5). About monthly samplings in the fishway (third experiment, spring–autumn 2020), for the two main sampled species (*Luciobarbus bocagei* and *Alburnus alburnus*), there was a relevant proportion of fish that entered the fishway with downstream migration direction (Table 5.4).

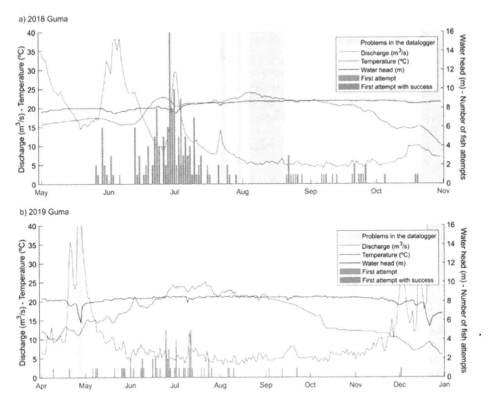

Fig. 5.4 Location and ascent success of Guma fishway, related to river discharge and water temperature. Water head refers to the difference between water level upstream and downstream. Between 30/10/2018 and 01/04/2019 data logger was running but there were no data at the antennas

5.2.4 Conclusions

Guma fishway is used not only for upstream migration but also for downstream migration and other types of movements related to the search of refuge or feeding purposes.

Fish were found using the fishway during most of the year, except in winter and their movements were mainly related to water temperature and peak discharge events.

The fishway does not seem to cause an important migration delay and it is suitable for upstream migration, at least for Iberian barbel.

Origin and length of the fish were identified as relevant factors. Fish from upstream origin showed a higher motivation for ascending and smaller fish showed a lower ascending success. A high degree of immature sizes was found using the fishway, thus, the lower success of small fish could be related to the absence of reproduction motivation.

Fig. 5.5 Main results of downstream migration trials in Guma fishway

Table 5.4 Proportion of fish moving downstream with respect to the total fish sampled in the fishway. Between brackets (number of fish descending / sum of fish descending and ascending)

Date	Luciobarbus bocagei	Alburnus alburnus
Jun-2020	58% (11/19)	85% (29/34)
Jul-2020	71% (12/17)	21% (28/132)
Aug-2020	35% (6/17)	25% (24/96)
Sep-2020	50% (21/42)	27% (23/84)
Oct-2020	100% (3/3)	100% (1/1)

5.3   Adaptation of Vertical Slot Fishways to Multi Fish Species with Macro-Roughness

5.3.1   Introduction

Vertical Slot Fishways (VSF) are technical solutions developed from more than 30 years for upstream fish migration. As they have been designed at first for salmonids who have high swimming capacities, they are not always suitable for other fish species. Adaptive technical solutions have been developed with the objective to facilitate the upstream migration for all the fish species and in particular for the small species (Albayrak et al. 2020).

5.3.2 Methods and Results

A basic Vertical Slot Fishway geometry is established (Fig. 5.6) by an analysis of the geometry of VSFs built over the last 20 years, from which we defined an 'average' geometry.

The main geometric ratios of the pool are the L/b which is fixed to 10 with L the length of the pool and b the width of the slot, B/b which varies from 9 to 5.67 with B the pool width. The baffles in the pools have the following ratios: A/b = 2 and a/b = 1.3. To ensure the migration of individuals or species with low swimming ability, these fishways can be improved by inserting elements such as vertical cylinders, sills in the slot, bottom roughness or flexible structures, which were suggested by Tarrade et al. (2011), Calluaud et al. (2014) or Ballu et al. (2019). The influence of added elements on the flow features is highlighted by laboratory and field experiments and flow numerical simulations. They were carried out for several configurations and for a wide range of channel slopes, pool widths and flow discharges. The modification of the flow topology, discharge coefficients and turbulence features were evaluated by ADV and water depth measurements. The results presented herein may be used to define a predictive law that helps engineers ensure the greatest effectiveness of VSFs with the presence of elements (Ballu et al. 2019). For example, macro-roughness are more and more often positioned at the bottom of fishways and they can facilitate the passage of benthic species through the crossing device (Branco et al. 2015). These are useful to reduce bottom velocity and create hydraulic shelters. However, the presence of macro-roughness is not yet taken into account when designing a VSF.

The fish migration efficiency varies with the flow topologies which are described in details in Calluaud et al. (2014). In particular, two salient flow topologies can be generated depending on the length to width ratios of the pools and on the slope, called the first flow pattern (FP1) and the second flow pattern (FP2). The FP1 provides better efficiency for

Fig. 5.6 Pool configuration and axes of references (Z is the vertical axis)

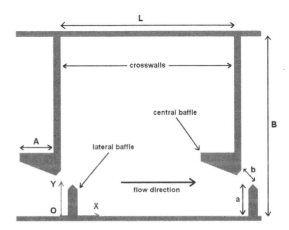

Fig. 5.7 Mapping of the transition zones between the type 1 topology and the type 2 topology as a function of the macro-roughness density

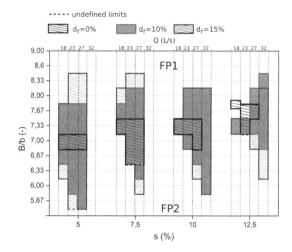

the upstream migration and is preferred for the design of a VSF. A mapping of the flow topology (Fig. 5.7) as a function of the pool width, the slope s expressed in %, and discharge is given for three densities of macro-roughness's, noted d_r (d_r is the surface occupied by the macro-roughness divided by the surface of the bottom of the pool). The density d_r is defined as the ratio between the area covered by elements with macro roughness and the total area on which the elements are positioned.

For the three macro-roughness densities, the graph of Fig. 5.7 delimits the transition zones of flow pattern for each parameter. It can be seen that for all configurations, an increase in the slope of the fishway forces the flow to adopt a FP2. In agreement with Wang et al. (2010) the flow discharge has no significant impact on the flow topology for $d_r = 0\%$. Since macro-roughness are introduced at the bottom of the fishway, the transition between FP1 and FP2 spreads over a wider range of pool width. This effect is amplified when the density of macro-roughness increases, where the change from FP1 to transition occurs at a greater B/b and the change from transition to FP2 at a smaller B/b. For instance, for Q = 23 l/s and s = 5%, the flow changes from FP1 to transition at B/b = 7 for $d_r = 0\%$, at B/b = 7.67 for $d_r = 10\%$, and finally at B/b = 8.33 for $d_r = 15\%$. Likewise, the flow changes to FP2 for B/b = 6.67 for $d_r = 0\%$, B/b = 5.67 for $d_r = 10\%$ and B/b smaller than 5.67 for $d_r = 15\%$. The flow discharge has an important role with $d_r = 10\%$ and $d_r = 15\%$, because it increases also the transition area between the two Flow Patterns. As the slope grows, the flow shifts from FP1 to transition for higher values of B/b when $d_r = 0\%$ and $d_r = 10\%$. However, this trend tends to be reversed when $d_r = 15\%$.

The influences of discharge, pool slope, pool width and density of macro-roughness on the discharge coefficient C_d are illustrated in Fig. 5.8. The discharge coefficient C_d (−) is the ratio between the measured discharge Q (m^3/s) and the theoretical discharge, which depends on the water depth at the centre of the pool and is expressed as follows,

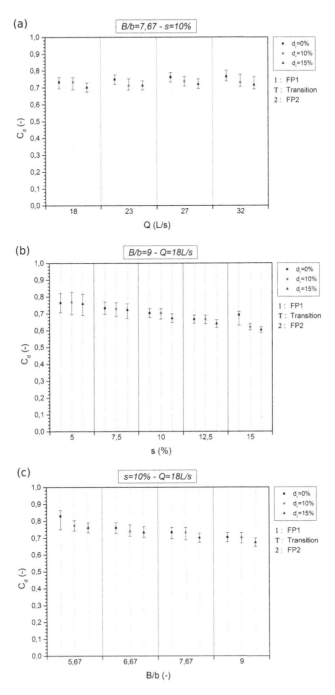

Fig. 5.8 Evolution of the discharge coefficient C_d with: **a** the flow discharge, **b** the slope and **c** the pool width

where g is the acceleration due to gravity (ms^{-2}), b is the width of the slot (m), h$_0$ is the water depth (m) at the centre of the pool, Δh the head difference (m) between successive basins, and Q is the discharge (m^3/s).

$$C_d = \frac{Q}{Q_{th}} = \frac{Q}{b.h_0.\sqrt{2.g.\Delta h}}$$

The uncertainty is reported on each plot using error bars. The mean value of relative uncertainty on the C_d is about 7% with a coverage factor k = 2.

Figure 5.8a shows that the value of C_d is not significantly affected by the change in Q. The biggest variation of C_d with the Q is measured when d$_r$ = 0% and is about 0,03 ±0,1 with k = 2. In contrast, an increase in the slope creates a significant decrease of C_d, with a minimum of −20% ±7% with k = 2 and d$_r$ = 15% (Fig. 5.8b). As regards to the influence of B/b (Fig. 5.8c), there is also a significant decrease of C_d when the pools become wider. The reduction of the discharge coefficient between B/b = 5.67 and B/b = 9 is about −15% ±8% with k = 2.

The mean evolution of C_d values between s = 5% and s = 12.5% was calculated for each of the four widths studied. With macro-roughness, an increase in pool width reveals a more important influence of the slope on the discharge coefficient value. For example, for B/b = 5.67 the relative decrease in C_d between s = 5% and s = 12.5% is about 4% for d$_r$ = 10% and 15% against more than 12% for B/b = 9. Without macro-roughness, it remains on average around 10%. Consequently, in the presence of macro-roughness and for small pool widths, the discharge coefficient will hardly be affected by a change in the channel slope.

To quantify the influence of macro-roughness density on the kinematic quantities of the flow, a comparison of the velocity and turbulent kinetic energy profiles (TKE) obtained for a pool width B/b = 6.67, a slope s = 7.5% and a flow rate Q = 23L/s in the without macro-roughness configuration, d$_r$ = 10% and d$_r$ = 15% is performed. When B/b = 6.67, the topology is a transient type in the three configurations (smooth floor, d$_r$ = 10% and 15%). The graphs in Fig. 5.9 show the values of the velocity norm and turbulent kinetic energy $\frac{k_{3D}}{V_d^2}$, for one transversal profile at mid depth in the configuration B/b = 6.67.

The presence of macro-roughness induces a decrease of the Turbulent Kinetic Energy k_{3D}/V_d^2. For the two macro-roughness densities, between Y/b = 0 et Y/b = 2, k_{3D}/V_d^2 is reduced by an average of 40% ±7% compared to the smooth floor configuration. On the 4 ≤Y/b≤6 interval, this reduction is estimated at an average 26% ±7%. It should be noted that on the first interval, it is the streamwise component k_U/V_d^2 which mainly contributes to the decrease in the global Turbulent Kinetic Energy k_{3D}/V_d^2, while it is equally distributed between k_V/V_d^2 et k_W/V_d^2 on the second interval. On the other hand, the impact of increased density of macro-roughness is negligible.

For this configuration (B/b = 6,67), the flow is transient regardless of the macro-roughness density. The area near the wall opposite the slot (4 $\leq \frac{Y}{b} \leq$ 6) is then subject to large velocity variations: it alternates between type 1 flow and type 2 flow. The x

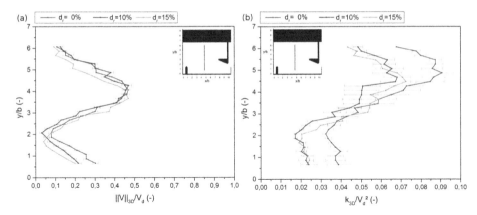

Fig. 5.9 Three-dimensional mean velocity $\frac{\|V_{3D}\|}{V_d}$ and Turbulent Kinetic Energy $\frac{k_{3D}}{V_d{}^2}$ according to the macro-roughness densities for a pool width of B/b = 6.67

component of the velocity is the more impacted by these variations, because this part of the profile is located in the jet in type 2 flow (highly positive velocity) and in the recirculation zone in the case of type 1 flow (negative velocity). With macro-roughness, the amplitude of these variations seems reduced.

The lower part of the profile ($0 \leq Y/b \leq 2$) is located in a recirculation zone. The presence of macro-roughness seems to create a constraint, independent of density, which decreases the amplitude of vertical and transverse velocity fluctuations.

The turbulent kinetic energy $\frac{k_{3D}}{V_d{}^2}$ is reduced by an average of $25\% \pm 15\%$. Areas of strong influence are mostly located on either side of the jet ($Y/b \leq 2$ et ≥ 4) and the three fluctuating components contribute equally to this decrease.

5.3.3 Conclusions

Vertical Slot Fishways (VSF) have been studied in this work to modify them and allow the upstream passage of not only salmonids but also species with small swimming capacities. Adaptive technical solutions have been developed and tested like the adjunction of one or three cylinders in the pool, sills in the slots, macro-roughness in the bottom of the pool or flexible cylinders inside the pool. The technical solutions proposed have shown some large modifications of the hydraulics and in particular the flow quantities which govern the fish motion (Velocity, Velocity Gradient, Turbulent Kinetic Energy and Dissipation), are reduced highly with the adjunction of elements. Values of the discharge coefficient for these different configurations are given to help the engineers to design such vertical slot fishways. A methodology is proposed which defines first the topology of the flow from the geometric parameters and finally predicts the discharge coefficients, (Ballu et al. 2019).

These different options could be used to modify existing, poorly functioning fishways and could allow the upstream migration of a greater number of fish species, at a lower cost than building a completely new fishway.

Acknowledgments We specifically thank Juan Carlos Romeral de la Puente (SAVASA) for the availability of the HPP facilities and for his continuous support, as well as Confederación Hidrográfica del Duero (Duero Water Authority) and Fishing Service of the regional government of Castilla y León for their legal and technical support.

References

Albayrak I, Boes R, Beck C, Meister J, David L, Lemkecher F, Chatellier L, Courret D, Pineau G, Calluaud D, Larrieu T, Sagnes P, Geiger F, Rutschmann P (2020) D3.4 – Enhancing and customizing technical solutions for fish migration. FIThydro Project Report, https://www.fithydro. eu/deliverables-tech/

Ballu A, Calluaud D, Pineau G, David L (2019) Experimental-based methodology to improve the design of Vertical Slot Fishways. J Hydraul Eng 145(9):04019031. https://doi.org/10.1061/(ASC E)HY.1943-7900.0001621

Benitez JP, Dierckx A, Nzau Matondo B, Rollin X, Ovidio M (2018) Movement behaviours of potamodromous fish within a large anthropised river after the reestablishment of the longitudinal connectivity. Fish Res 207:140–149

Branco P, Amaral SD, Ferreira MT, Santos JM (2017) Do small barriers affect the movement of freshwater fish by increasing residency? Sci Total Environ 581:486–494. https://doi.org/10.1016/ j.scitotenv.2016.12.156

Bravo-Córdoba FJ, Sanz-Ronda FJ, Ruiz-Legazpi J, Valbuena-Castro J, Makrakis S (2018) Vertical slot versus submerged notch with bottom orifice: Looking for the best technical fishway type for Mediterranean barbels. Ecol Eng 122:120–125. https://doi.org/10.1016/j.ecoleng.2018.07.019

Bunt CM, Castro-Santos T, Haro A (2012) Performance of fish passage structures at upstream barriers to migration. River Res Appl 28(4):457–478

Calluaud D, Pineau G, Texier A, David L (2014) Modification of vertical slot fishway flow with a supplementary cylinder. J Hydraul Res 52(5):614–629

Dingle H (2014) Migration. Oxford University Press, Oxford, The biology of life on the move, p 326

Illies J, Botoseanu L (1963) Problèmes et méthodes de la classification et de la-zonation écologique des eaux courantes, considérées surtout-du point de vue faunistique. SIL Commun 12:1–57. https://doi.org/10.1080/05384680.1963.11903811

IUCN (2018) The IUCN red list of threatened species (www.iucnredlist.org). [WWW Document]

Lucas MC, Baras E (2001) Migration of freshwater fishes. Blackwell Science, London, p 420

Nilsson C, Reidy CA, Dynesius M, Revenga C (2005) Fragmentation and flow regulation of the world's large river systems. Science 308(5720):405–408. https://doi.org/10.1126/science.110 7887

Noonan MJ, Grant JWA, Jackson CD (2012) A quantitative assessment of fish passage efficiency. Fish Fish 13(4):450–464

Peter A, Mettler R, Schölzel N (2016) Kurzbericht zum Vorprojekt "PIT-Tagging Untersuchungen am Hochrhein – Kraftwerk Rheinfelden", p 45

Rosgen DL, Silvey HL (1996) Applied river morphology. Wildland Hydrology, Pagosa Springs, Colorado, USA

Tarrade L, Pineau G, Calluaud D, Texier A, David L, Larinier M (2011) Detailed experimental study of hydrodynamic turbulent flows generated in vertical slot fishways. J Environ Fluid Mech 11(1):1–21

Wang R, David L, Larinier M (2010) Contribution of experimental fluid mechanics to the design of vertical slot fish passes. Knowl Manag Aquat Ecosyst 396:02

Wootton RJ (1990) Ecology of teleost fishes. Chapman & Hall, London, p 404

Attraction Flow and Migration Habitat Assessment Using an Agent-Based Model

6

Ianina Kopecki, Matthias Schneider⊙, and Tobias Hägele

6.1 Introduction

One important key to re-establish sustainable fish populations in rivers is the fish habitat connectivity. In most rivers, the connectivity is disturbed through multiple obstructions such as hydropower plants, weirs and sills. Even if those are supplied with well-functioning fish ladders, the detectability of these facilities for the migratory fish usually plays a critical role in the overall passability of a river barrage. Attraction flow is one of the major factors defining fishway performance (Bunt et al. 2011; Cooke and Hinch 2013; Silva et al. 2018). In many cases investigations focus on the attraction flow rate, i.e., the proportion of the flow from the fishpass to the flow from the adjacent turbines or a weir (Larinier et al. 2008). The flow velocity magnitude in the vicinity of the fishpass outlet and its rate in comparison to the surrounding river flow velocity is another parameter often considered (Williams et al. 2012). Various studies indicate that other factors can influence fish movement when approaching a fishpass entrance. Turbulence (Liao 2007; Kirk et al. 2017), location of the attraction flow outlet (Burnett et al. 2016), and other physical and chemical parameters such as for example temperature (Capra et al. 2017; Caudill et al. (2013) Indirect effects of impoundment on migrating fish: temperature gradients in fish ladders slow dam passage by adult Chinook salmon and steelhead. PloS One

I. Kopecki · M. Schneider (✉) · T. Hägele
sje – Ecohydraulic Engineering GmbH, Stuttgart (Vaihingen), Germany
e-mail: mailbox@sjeweb.de

I. Kopecki
e-mail: kopecki@sjeweb.de

T. Hägele
e-mail: haegele@sjeweb.de

P. Rutschmann et al. (eds.), *Novel Developments for Sustainable Hydropower*,
https://doi.org/10.1007/978-3-030-99138-8_6

8(12), e85586) or light and noise (Popper and Carlson 1998) have been studied and could potentially affect migration. Other factors like water depth (Scheibe and Richmond 2002, Goodwin et al. 2006), river morphology and obstacles on the river bottom may influence fish movements as well (Piper et al. 2012 and 2015). Within the FIThydro project the habitat model CASiMiR (Noack et al. 2013) has been extended by an agent-based module. The model aims on the assessment of probability for fish being routed into the outlet of a fishway.

6.2   Test Case Altusried

The model approach has been developed and evaluated using fish track data collected at the test case site Altusried hydropower plant (HPP) with an acoustic telemetry system. The HPP Altusried is one of 5 HPPs in the Upper Iller River, a tributary of River Danube in South-West Germany (Fig. 6.1, left). It is in operation since 1961, has a hydraulic head of 3 m, an installed capacity of 1,6 MW, the maximum turbine flow is 80 m^3/s and it is equipped with 2 Kaplan turbines. They are located at the left side of the weir. All 5 HPPs in the Upper Iller River are operated by the Bayerische Elektrizitätswerke (BEW) and have lately been equipped with facilities for upstream migration. The fishpass outlet in Altusried is located about 260 m below the weir (Fig. 6.1, right).

The telemetry network consisted of 16 receivers 180 kHz HR2 (High Residency) with built-in synchronization tags and temperature, noise and tilt sensors, and 6 reference tags.

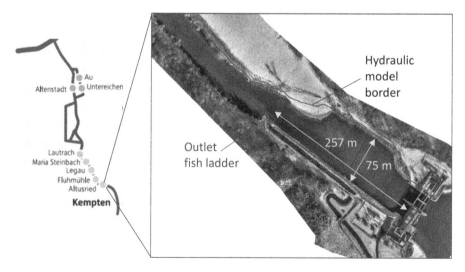

Fig. 6.1 Location of HPP Altusried (left) and an aerial picture showing the HPP and the location the fish ladder and its outlet (right)

In total, 25 grayling and 22 barbel were caught, tagged and released in the telemetry system array during their respective migration period in spring 2019.

6.3   Migration Model Concept

The concept of the migration model combines habitat suitability maps for migrating fish with additional information on the swimming behaviour of the observed fish in the flow field. Hydrodynamic parameters that define migrating corridors are derived upon the statistical analysis of fish tracks of European grayling (*Thymallus thymallus*) and barbel (*Barbus barbus*) recorded over the spring period of 2019 downstream of the HPP Altusried. Swimming behaviour is expressed in terms of a histogram of the probability of fish to change the swimming direction compared to the one in the previous movement step.

The present model operates on the results of a 2D hydrodynamic model. Flow velocity vectors, velocity magnitude and flow depth, obtained with the model Hydro_AS-2D (Nujic 2006) on an unstructured mesh, are interpolated to a structured grid of the migration model. Starting at a defined specific location, a virtual fish, the so-called *fish-agent*, evaluates the current surrounding flow field and selects the most probable next movement direction according to the pre-defined behavioural rules. Step by step, the path of a fish-agent is calculated and visualized, allowing the modeler to get a picture of possible migratory movements in the far vicinity of a fish ladder outlet. The following main elements form the basis of the migration algorithm:

- Fish agents in their search for the upstream path swim within the so-called *migration corridors* defined by migration habitat suitability. For the demarcation of migration corridors, a CASiMiR fuzzy rule-based approach is applied (Noack et al. 2013). Parameters defining these corridors and corresponding fuzzy rules and sets are detected through the analysis of the observed fish positions prior entering the fish migration facility. A map of the output parameter "Migration corridor suitability" shows which parts of the river are preferable for migration (see example of migration corridor in Fig. 6.4). Swimming along the migration corridor, a virtual fish is assumed to prefer locations with higher suitability and move less likely into the locations with lower suitability compared to the suitability in the current position. Fish-agent's moves depend on flow direction and are allowed only in the areas with velocities within the rheoreaction thresholds (e. g. restricted to the regions with flow velocities in the range between the rheotactic detectability threshold and burst swimming speed of the target fish species (see e. g. Adam and Lehmann 2011). Orientation in the flow field and selection of the next movement direction is chosen upon the histogram of probability to move within the flow field. This histogram is obtained upon the analysis of fish movements 30 min prior to the first entry into the fishpass. It describes the observed behaviour of fish deviating from a straight path while moving towards the entrance of fishpass.

6.4 Fuzzy Systems for Migration Corridors

Two fuzzy rule-based model versions were tested for the calculation of the migration suitability: One with two parameters (flow velocity and water depth) and the other with four parameters (flow velocity, water depth and spatial gradients of flow velocity and water depth). The comparison of recorded fish tracks and calculated hydraulic parameters shows that in the final migration phase approaching the fishway, many fish individuals move along the areas where hydraulic parameters change abruptly. Thus, it can be expected that spatial gradients play an important role for fish migration. However, for brevity reasons, only the two-parameter system based on flow velocity and water depth is presented here.

Both fuzzy rule systems are derived through the analysis of the observed fish tracks in the time-period of 30 min prior to the first entry into the fishpass. Figure 6.2 (right) shows the frequency distribution (blue bars) of the four hydraulic parameters for all observed grayling during the above-mentioned time-period. Based on these frequency distributions of spatial use, up to five fuzzy sets (categories that indicate the preference of fish to use

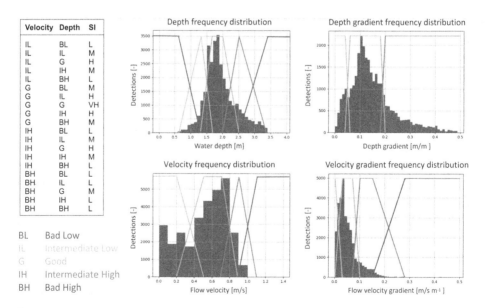

Fig. 6.2 Grayling: fuzzy rules (left) for the two parameter system, and fuzzy sets for the four parameter system: Water depth (centre top), flow velocity (centre bottom), depth gradient (right top), velocity gradient (right bottom)

certain hydraulic conditions) are derived (see Fig. 6.2). Combining those fuzzy sets with fuzzy rules (Fig. 6.2, left), the migration corridor suitability maps are calculated.

Rule 1 example:	IF flow velocity is Intermediate Low AND water depth is Bad Low THEN migration corridor suitability is Low
Rule 2 example:	IF flow velocity is Intermediate Low AND water depth is Intermediate Low THEN migration corridor suitability is Medium

6.5   Model Results

Some simulation results are presented in the following figures. They are overlaid with the ortho images of the HPP Altusried. The comparison of observed tracks with modelled tracks (example for grayling in Fig. 6.3) shows that the concept of flow probability histogram combined with the random method for the final selection of movement direction allows to mimic up to a certain degree the searching behaviour of observed fish (random lateral, zig-zag). However, individual fish paths of all tagged fish vary widely and many movements are most probably not directly related to migration behavior.

6.6   Analysis of Migratory Situation—Mitigation Options

The benefit of the simulation model is that it allows to demarcate migration corridors indicating areas most probable to be used for upstream migration and fish migration paths for different flow scenarios. This makes it a suitable tool to develop mitigation options for the detectability of fishway entrances.

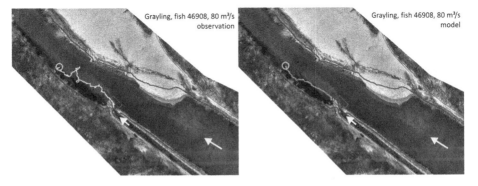

Fig. 6.3 Observation and migration model results for grayling at 80 m^3/s: Fish 46,908 observation (left), Fish 46,908 model with 2 parameters migration fuzzy system (right)

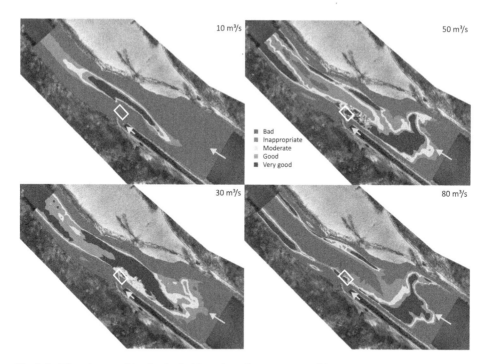

Fig. 6.4 Migration corridor for barbel based on fuzzy system with two parameters (flow velocity and water depth) and its change with increasing HPP discharge: 10 m³/s (top left), 30 m³/s (bottom left), 50 m³/s (top right) and 80 m³/s (bottom right), entrance area marked by a white rectangle

Options to mitigate the attraction flow can be for example a seasonal adaptation of the turbine flow from a considered HPP or release of additional water into the river in direct neighbourhood of the fishpass outlet. Simulation scenarios in Fig. 6.4 show that migration corridors for barbel are getting narrower with increasing discharge from the HPP but at the same time are shifting from the middle of the river towards the banks, which could increase the chance for fish to find the fishpass outlet. In contrast, increasing the discharge too much, up to 80 m³/s and higher, leads to an interruption of the migration corridor on the left river side, which could impair the attraction to the fishpass outlet (Fig. 6.4, bottom right).

6.7 Outlook

First results of an agent-based model are promising. The current model mimics the movement of individual fish by a combination of migration habitat suitability maps with behavioural rules of fish derived from observations of fish movements in the flow field. The base parameters for the definition of migration corridors are flow velocity, water depth and hydraulic gradients. Predicted swimming paths of grayling and barbel show high similarities with observed tracks of individual fish.

The evaluation of the model runs for multiple fish-agents for the test site Altusried confirms a basically appropriate position of the fishpass entrance. HPP flow rates in the range from 40 to 50 m^3/s for barbel and from 40 to 80 m^3/s for grayling seem to be favourable in the migration season. Fish moving upstream along the left riverbank find the entrance with a much higher probability than those moving close to the right bank.

Further developments will concentrate on more detailed processing of fish tracks aiming to distinguish between different behaviour types (feeding, resting, searching). Additional investigations are planned for the final identification of key hydrodynamic and environmental parameters for the migration model.

Acknowledgements We specifically thank Ine Pauwels and her team from the Research Institute Nature and Forest (INBO) and Tobias Epple from the University of Augsburg and LEW Wasserkraft GmbH for the support in the fish telemetry study in the case study Altusried.

References

Adam B, Lehmann B (2011) Ethohydraulik. Springer-Verl, Berl Heidelb, Grundlagen, Methoden und Erkenntnisse

Bunt CM, Castro-Santos T, Haro A (2011) Performance of fish passage structures at upstream barriers to migration. River Res Applic. https://doi.org/10.1002/rra.1565:n/a

Burnett NJ, Hinch SG, Bett NN, Braun DC, Casselman MT, Cooke SJ, Gelchu A, Lingard S, Middleton CT, Minke-Martin V, White CFH (2016) Reducing carryover effects on the migration and spawning success of sockeye salmon through a management experiment of dam flows. River Res Appl. https://doi.org/10.1002/rra.3051

Capra H, Plichard L, Bergé J, Pella H, Ovidio M, McNeil E, Lamouroux N (2017) Fish habitat selection in a large hydropeaking river: Strong individual and temporal variations revealed by telemetry. Sci Total Environ 578:109–120. https://doi.org/10.1016/j.scitotenv.2016.10.155

Caudill CC, Keefer ML, Clabough TS, Naughton GP, Burke BJ, Peery CA (2013) Indirect effects of impoundment on migrating fish: temperature gradients in fish ladders slow dam passage by adult Chinook salmon and steelhead. PloS One 8(12):e85586

Cooke SJ, Hinch SG (2013) Improving the reliability of fishway attraction and passage efficiency estimates to inform fishway engineering, science, and practice. Ecol Eng, 58:123–132. https://doi.org/10.1016/j.ecoleng.2013.06.005

Goodwin RA, Nestler JM, Anderson JJ, Weber LJ, Loucks DP (2006) Forecasting 3-D fish movement behavior using a Eulerian-Lagrangian-agent method (ELAM). Ecol Model 192(1–2):197–223

Kirk MA, Caudill CC, Syms JC, Tonina D (2017) Context-dependent responses to turbulence for an anguilliform swimming fish, Pacific lamprey, during passage of an experimental vertical-slot weir. Ecol Eng 106:296–307. https://doi.org/10.1016/j.ecoleng.2017.05.046

Larinier M (2008) Fish passage experience at small-scale hydro-electric power plants in France. Hydrobiol 609:97–108

Liao JC (2007) A review of fish swimming mechanics and behaviour in altered flows. Philos Trans
    R Soc B: Biol Sci 362(1487):1973–1993. https://doi.org/10.1098/rstb.2007.2082
Noack M, Schneider M, Wieprecht S (2013) The Habitat Modelling System CASiMiR: A Multi-
    variate Fuzzy-Approach and its Applications. In: Maddock I, Harby A, Kemp P, Wood P (Eds.),
    Ecohydraulics: an integrated approach. John Wiley & Sons, 8/2013
Nujic MS (2006) Hydro_AS-2D. Ein zweidimensionales Strömungsmodell für die wasser-
    wirtschaftliche Praxis. Benutzerhandbuch
Piper AT, Manes C, Siniscalchi F, Marion A, Wright RM, Kemp PS (2015) Response of seaward-
    migrating european eel (Anguilla anguilla) to manipulated flow fields. Proc R Soc B: Biol Sci
    282(1811):1–9
Piper AT, Wright RM, Kemp PS (2012) The influence of attraction flow on upstream passage of
    European eel (Anguilla anguilla) at intertidal barriers. Ecol Eng, 44:329–336.https://doi.org/10.
    1016/j.ecoleng.2012.04.019
Popper AN, Carlson TJ (1998) Application of sound and other stimuli to control fish behav-
    ior. Trans Am Fish Soc 127(5):673–707. https://doi.org/10.1577/1548-8659(1998)127%3c0673:
    AOSAOS%3e2.0.CO;2
Scheibe TD, Richmond MC (2002) Fish individual-based Numerical Simulator (FINS): A particle-
    based model of juvenile salmonid movement and dissolved gas exposure history in the Columbia
    River Basin. Ecol Model 147(3):233–252
Silva AT, Lucas MC, Castro-Santos T, Katopodis C, Baumgartner LJ, Thiem JD, Aarestrup K, Pom-
    peu PS, O'Brien GC, Braun DC, Burnett NJ, Zhu DZ, Fjeldstad HP, Forseth T, Rajaratnam N,
    Williams JG, Cooke SJ (2018) The future of fish passage science, engineering, and practice. Fish
    Fish 19(2):340–362. https://doi.org/10.1111/faf.12258
Williams JG, Armstrong G, Katopodis C, Larinier M, Travade F (2012) Thinking like a fish: A key
    ingredient for development of effective fish passage facilities at river obstructions. River Res
    Appl 28(4):407–417

Fish Guidance Structures with Narrow Bar Spacing: Physical Barriers

7

Laurent David, Ludovic Chatellier, Dominique Courret, Ismail Albayrak, and Robert M. Boes

7.1 Introduction

Fish migration in regulated rivers is often hampered by Hydropower plants (HPPs), dams, weirs and spillways. The main risks associated with the presence of such structures include: blocking or delaying of up- and downstream fish migrations, and damage or mortality of fish when passing turbines, weirs or spillways. For an efficient restoration of water bodies, the European Water Framework Directive was enacted in 2000 and the revised Swiss Waters Protection Act (WPA) and Waters Protection Ordinance were introduced in 2011. For HPPs, fish passage facilities and connections to adjoining water bodies must be upgraded or newly erected.

L. David (✉) · L. Chatellier
Institut Pprime, Pole Ecohydraulique OFB/IMFT/Pprime, CNRS University of Poitiers, Poitiers, France
e-mail: laurent.david@univ-poitiers.fr

L. Chatellier
e-mail: ludovic.chatellier@univ-poitiers.fr

D. Courret
OFB/IMFT/Pprime, Office Français de La Biodiversité, Pole ecohydraulique, Toulouse, France
e-mail: dominique.courret@imft.fr

I. Albayrak · R. M. Boes
Laboratory of Hydraulics, Hydrology and Glaciology, ETH Zurich, Zurich, Switzerland
e-mail: albayrak@vaw.baug.ethz.ch

R. M. Boes
e-mail: boes@vaw.baug.ethz.ch

© The Author(s) 2022
P. Rutschmann et al. (eds.), *Novel Developments for Sustainable Hydropower*,
https://doi.org/10.1007/978-3-030-99138-8_7

Downstream fish passage still poses challenges to scientists, engineers, authorities and HPP operators due to the lack of design standards and related basic information on behaviour of various fish species. To this end, FIThydro has improved and developed downstream fish passage technologies for a range of fish species and size of HPPs based on laboratory and field investigations. These are classified into two groups, namely fish guidance structures with narrow bar spacing (i) and wide bar spacing (ii). The former group includes vertically inclined bar rack-bypass system and horizontal bar rack-bypass system, which work as a physical barrier and potentially applied for small to medium size HPPs with a design discharge <120 m^3/s and this chapter deals with the hydraulics, fish guidance efficiency and design recommendations of this type of fish guidance structures. The latter group includes an innovative curved-bar rack-bypass system, which functions as a combination of mechanical behavioural barrier for small-to-large HPPs as addressed in Chap. 8.

7.2   Physical Barriers

Different solutions have been widely studied in multiple designs (inclined, angled, vertical or horizontal bars, different bar shapes and bar spacings) during the FIThydro project. They proved their efficiency in avoiding fish passage through and impingement risks at the rack and additionally guide fishes towards a downstream bypass. However, narrow bar spacing racks cause problems of head losses and clogging by floating debris such as leaves and wood, requiring efficient cleaning systems and generating additional maintenance costs compared to classical intake trash racks. In this sub chapter, different solutions with narrow bar spacing are explained and summarized.

The first type of such racks is the Vertically Inclined Bar Racks (VIBR). They consist of plane screens composed of elongated flat bars positioned in vertical planes aligned with the flow (Fig. 7.1). The plane screen is inclined with an angle β with respect to the river bed to guide fish towards one or several surface bypass inlets located at the top of the rack (Raynal et al. 2013a). Another configuration consists of a perforated plate instead of bars, which is called Vertically Inclined Perforated Plate (VIPP). Detailed information on the design and efficiency of both VIBR and VIPP is given in FIThydro Deliverable 3.4 (Albayrak et al. 2020) and Lemkecher (2020).

The second type of solutions are angled bar racks. Angled bar racks are installed at an angle α to the approach flow direction in plan view to guide fish towards a bypass located at the downstream end of the rack. Three types of angled racks with narrow bar spacing, $s_b \leq 30$ mm, can be distinguished (Fig. 7.2):

"Classical" angled bar rack, with vertical bars angled (Raynal et al. 2013b)
Angled bar rack with vertical bars oriented in streamwise direction (Raynal et al. 2014)

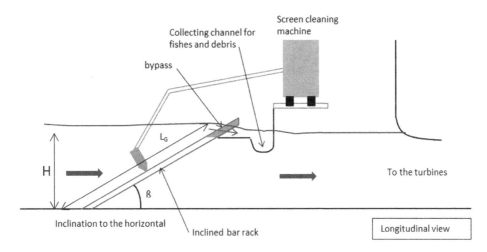

Fig. 7.1 Longitudinal profile of a vertically inclined bar rack (from Courret and Larinier 2008)

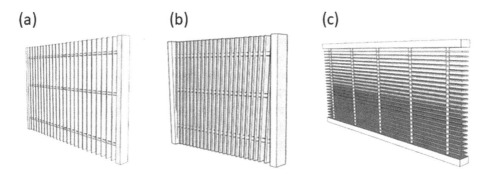

Fig. 7.2 Type of fish guidance structures with narrow bar spacing: angled bar rack with vertical bars (**a**), vertical streamwise bars (**b**) and horizontal bars (**c**) (adapted from Lemkecher et al. 2022)

Horizontal Bar Rack (HBR)' (Albayrak et al. 2019, 2020, Meister 2020; Meister et al. 2020a, b; Lemkecher et al. 2022)

Figure 7.3 shows the horizontal bar rack—bypass system (HBR-BS) of the FIThydro case study residual HPP Schiffmühle on the Limmat River, Switzerland, during revision work in 2018. The design discharge of the HPP is $Q_d = 14$ m^3/s and the HBR was built in 2013 with foil-shaped bars, a clear bar spacing of $s_b = 20$ mm, and a pipe bypass.

Inclined bar racks (VIBRs) and horizontal bar racks (HBRs) are characterized by narrow bar spacing typically ranging between $s_b = 10$ and 30 mm, such that they are physically not passable for a large share of the fish population. VIBRs and HBRs are thus designed as physical fish exclusion and guidance barriers to prevent fish from entering

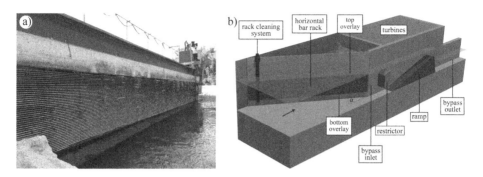

Fig. 7.3 **a** Horizontal bar rack—bypass system at the residual flow HPP Schiffmühle, Switzerland, during revision work in July 2018. (*Source* Julian Meister, VAW) and **b** principle sketch of an HBR-BS (*Source* VAW, adapted from Ebel 2016)

water intakes or the turbines at run-of-river HPPs. As a rule of thumb, the rack constitutes a physical barrier when the bar spacing is lower than 1/10 of the total length for most species including salmonids, but except for eels, which require bar spacing lower than 3% of their length (Ebel 2016). For fish smaller than the threshold size, VIBRs and HBRs act as behavioural barriers. The lower the bar spacing, the higher the fish will be reluctant to go through the rack.

Bottom and top overlays can be used to enhance the guidance efficiency of sediments, floating debris, and bottom and surface oriented fish, respectively Fig. 7.3. An automated rack cleaning machine is needed to prevent the rack from clogging. In case of HBRs, Fig. 7.3 illustrates that the bypass discharge is usually controlled with a restrictor and/or a ramp. In case of VIBRs, Fig. 7.4 illustrates that one or several bypass entrances, depending on the intake width, are collected in transversal galleries with growing hydraulic sections in the downstream direction.

The bars of VIBRs and HBRs can be built with different bar shapes, such as rectangular, rectangular with a circular tip, rectangular with an ellipsoidal tip & tail, and foil-shaped (Fig. 7.5). Most modern VIBRs and HBRs are equipped with foil-shaped bars or rectangular bars with an ellipsoidal tip & tail because of the reduced head losses (Lemkecher et al. 2020; Meister et al. 2020a). Additionally, these bars can be cleaned more easily than rectangular bars due to the thickness reduction from tip to tail (Meister 2020). Figure 7.5 shows the different rack parameters of an HBR, including the clear bar spacing s_b, the bar thickness t_b, and the bar depth d_b (see Albayrak et al. 2020 for more information on HBR-BS).

To prevent fish from passing through the Flow Guidance Structure (FGS) with narrow bar spacing, there are three design criteria: (i) the bar spacing, (ii) the normal velocity (V_n; velocity component normal to the rack axis), which is directly linked to the rack surface, and (iii) the ratio of the rack parallel velocity (V_p) to the rack normal velocity,

Fig. 7.4 Views of VIBR installed at Las Rives HPP, France: inclined rack and bypass entrances viewed from upstream (**a**), (water intake out of water) and gallery collecting the 3 surface bypass entrances (**b**)

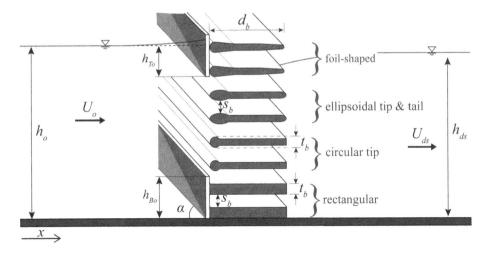

Fig. 7.5 Cross-section of an HBR illustrating different rack parameters; h_o: approach flow depth, h_{ds}: downstream flow depth, U_o: mean approach flow velocity, U_{ds}: mean downstream flow velocity, h_{Bo}: bottom overlay height, h_{To}: top overlay height, s_b: clear bar spacing, t_b: bar thickness at thickest point, d_b: bar depth. (*Source* VAW, adapted from Meister et al. (2020a))

which should be higher than 1 or even 2, i.e. $V_p/V_n > 1$ (or 2). The maximum values of the first two parameters depend on the species taken into account.

The recommended bar spacing and normal velocity (V_n) are the same for inclined racks (VIBR), angled racks with horizontal bars (HBR) and with vertical streamwise bars, as the behavioural "louver effect" is not considered strong enough in such configuration.

For salmonid smolts, the bar spacing (for inclined and angled bar racks) has to be smaller than 10–15 mm to constitute a physical barrier (based on the rule 1/10 of body width), but a strong behavioral repulsion can be obtained with bar spacing up to 25 mm (Courret and Larinier 2008). As eels do not show strong behavioural repulsion and are therefore likely to pass through trash racks, it appeared necessary to implement physical barriers. In France, the recommended bar spacing (for inclined and angled bar racks) is generally 20 mm to stop female eels longer than 50–60 cm. The bar spacing can be reduced to 15 mm in case of a significant presence of males upstream of the HPP (Courret and Larinier 2008). In Germany, the authorities in some regions even go below these values, with prescribed thresholds down to 10–12 mm.

For HBRs, the horizontal approach flow angle α, is selected such that the velocity component normal to the rack V_n does not exceed the sustained swimming speed of the target fish species. Approach flow velocities, typically varying between $U_o = 0.40$ and 0.80 m/s, lead to $\alpha = 20 \div 40°$. The rack angle is therefore a compromise between limiting V_n on the one hand and the rack length on the other hand. For Vertically Inclined Bar Racks (VIBRs), approach velocities have to respect the same criteria as the HBRs regarding V_n and rack inclinations of the order of 25° are necessary to guide fish towards surface bypass entrances—thus confirming existing recommendations ($V_p/V_n > 2$)—and helping to limit head losses (Courret and Larinier 2008; Courret et al. 2015).

The head losses induced by HBRs can be predicted with the equations published in Meister et al. (2020a) and Lemkecher et al. (2022). These equations do not only take rack parameters, as defined in Fig. 7.5, into account, but also different approach flow configurations as determined by the HPP layout such as diversion HPP or block-type HPP. If an HBR is installed in a straight headrace channel of a diversion HPP, the velocities are typically nearly homogeneously distributed, which means that the criterion of $V_p/V_n > 1$ is fulfilled for HBRs with $\alpha < 45°$ (Meister et al. 2020b). If an HBR is installed at a block-type HPP, the streamline pattern is usually complex and V_p/V_n along the rack decreases towards the downstream rack end (Meister et al. 2020b). Likewise, V_n will be underestimated at the downstream rack end if the velocity components are calculated from continuity, which could lead to fish impingements or passages through the rack. It is therefore recommended to determine the optimal HBR position with numerical simulations such as described in Feigenwinter et al. (2019).

The head losses of VIBRs and VIPP can be predicted using the equations developed by Lemkecher (2020).

In addition to the design of a FGS with narrow bar spacing, the bypass design is important to safely collect and transport the fish and to return them unharmed to the river downstream of an HPP. Different bypass designs are described in literature such as the full depth open channel bypass, a bypass with a vertical axis gate consisting of bottom and top openings, and a pipe bypass (Beck 2020; Meister 2020). The latter is not recommended because it can clog easily and fish avoid large velocity gradients at the inlet of the pipe bypass (Dewitte and David 2019).

The height and the width of the turbine intake influence the choice of the solution (inclined or angled). In addition, the possible location of the bypasses could modify the final solution. To reduce head losses, a particular attention has to be paid on the bar shape, the spacers and the support structures of the bar rack. For more details, please see the FIThydro Deliverables 2.2 (Dewitte and David 2019) and 3.4 (Albayrak et al. 2020); and the FIThydro Wiki on FGSs with narrow bar openings.

References

Albayrak I, Boes R, Beck C, Meister J, David L, Lemkecher F, Chatellier L, Courret D, Pineau G, Calluaud D, Larrieu T, Sagnes P, Geiger F, Rutschmann P (2020) D3.4 – Enhancing and customizing technical solutions for fish migration. FIThydro Proj Rep. https://www.fithydro.eu/del iverables-tech/

Beck C (2020) Fish protection and fish guidance at water intakes using innovative curved-bar rack bypass systems. VAW-Mitteilung 257 (R.M. Boes, ed). VAW, ETH Zurich, Switzerland. https://vaw.ethz.ch/en/the-institute/publications/vaw-communications/2010-2019.html

Courret D, Larinier M (2008) Guide pour la conception de prises d'eau 'ichtyocompatibles' pour les petites centrales hydroélectriques (Guide for the design of fish-friendly intakes for small hydropower plants). Agence de l'Environnement et de la Maîtrise de l'Energie (ADEME) (in French)

Courret D, Larinier M, David L, Chatellier L (2015, June 24) Development of criteria for the design and dimensioning of fish-friendly intakes for small hydropower plant. In: International conference on engineering and ecohydrology for fish passage, The Netherlands. Groningen. https://sch olarworks.umass.edu/fishpassage_conference/2015/June24/16/

Dewitte M, David L (2019) D2.2 – Working basis of solutions, models, tools and devices and identification of their application range on a regional and overall level to attain self-sustained fish populations. FIThydro Project Report. https://www.fithydro.eu/deliverables-tech/

Ebel G (2016) Fischschutz und Fischabstieg an Wasserkraftanlagen – Handbuch Rechen- und Bypasssysteme. Ingenieurbiologische Grundlagen, Modellierung und Prognose, Bemessung und Gestaltung (Fish Protection and Downstream Passage at Hydro Power Stations—Handbook of Bar Rack and Bypass Systems. Bioengineering Principles, Modelling and Prediction, Dimensioning and Design), 2nd ed.; Büro für Gewässerökologie und Fischereibiologie Dr. Ebel: Halle (Saale), Germany (In German)

Feigenwinter L, Vetsch DF, Kammerer S, Kriewitz CR, Boes RM (2019) Conceptual approach for positioning of fish guidance structures using CFD and expert knowledge. Sustain, 11(6):1646. https://doi.org/10.3390/su11061646

Lemkecher F (2020) Étude des grilles des prises d'eau ichtyocompatibles (Investigation of fish-friendly racks). Thesis Univ Poitiers, France. https://theses.univ-poitiers.fr/64422/2020-Lemkec her-Fatma-These

Lemkecher F, Chatellier L, Courret D, David L (2020) Contribution of different elements of inclined trash racks to head losses modelling. Water 12(966). https://doi.org/10.3390/w12040966

Lemkecher F, Chatellier L, Courret D, David L (2022) Experimental study of fish-friendly angled trash racks with horizontal bars. J Hydraul Res 60(1):136–147. https://doi.org/10.1080/002 21686.2021.1903587

Meister J (2020) Fish protection and guidance at water intakes with horizontal bar rack bypass systems. VAW-Mitteilung 258 (R.M. Boes, ed.). Laboratory of Hydraulics, Hydrology and Glaciology, ETH Zurich, Switzerland. https://ethz.ch/content/dam/ethz/special-interest/baug/vaw/vaw-dam/documents/das-institut/mitteilungen/2020-2029/258.pdf

Meister J, Fuchs H, Beck C, Albayrak I, Boes RM (2020a) Head losses of horizontal bar racks as fish guidance structures. Water 12(2):475. https://doi.org/10.3390/w12020475

Meister J, Fuchs H, Beck C, Albayrak I, Boes RM (2020b) Velocity Fields at Horizontal Bar Racks as Fish Guidance Structures. Water 12(1):280. https://doi.org/10.3390/w12010280

Raynal S, Châtellier L, Courret D, Larinier M, Laurent D (2013a) An experimental study on fish-friendly trashracks—Part 2. Angled trashracks. J Hydraul Res 51(1):67–75. https://doi.org/10.1080/00221686.2012.753647

Raynal S, Châtellier L, Courret D, Larinier M, Laurent D (2013b) An experimental study on fish-friendly trashracks—Part 1. Inclined trashracks. J Hydraul Res 51(1):56–66. https://doi.org/10.1080/00221686.2012.753646

Raynal S, Châtellier L, Courret D, Larinier M, David L (2014) Streamwise bars in angled trashracks for fish protection at water intakes. J Hydraul Res 52(3):426–431. https://doi.org/10.1080/002 21686.2013.879540

Fish Guidance Structure with Wide Bar Spacing: Mechanical Behavioural Barrier

8

Ismail Albayrak and Robert M. Boes

8.1 Introduction

Horizontally or vertically inclined FGS with narrow bar spacing of $s_b = 10$–30 mm described in Chap. 7 are not recommended for medium- to large-scale HPPs with a design discharge $Q_d > 100$ m^3/s because of their velocity limitations to avoid fish impingement (Ebel 2016) and relatively high clogging risk by floating debris and hence operational problems. For these HPPs, mechanical behavioural FGS with wide bar spacing of $s_b = 25$–100 mm present a promising alternative (Albayrak et al. 2018, 2020). They guide fish to a bypass with hydrodynamic cues created by the vertical bars instead of physically blocking fish from entering the water intake. When approaching the FGS, fish should perceive high turbulence zones and spatial velocity and pressure gradients around and between the bars and avoid passing the FGS. The velocity component parallel to the FGS guides fish towards the bypass located at the downstream end of the FGS. Louvers belong to this type of FGS with straight vertical bars placed normal to the approach flow, i.e., with a bar angle of $\beta = 90°$, and a rack angle to the approach flow of $\alpha = 10$–$45°$ (Amaral 2003, Bates and Vinsonhaler 1957, EPRI and DML 2001, Fig. 8.1a). They are widely used to bypass anadromous fish around HPPs and water intakes in the northeast USA and Canada. Furthermore, classical angled bar racks are also used for fish guidance similar to louvers, but their bars are placed at $90°$ to the rack axis, so that β varies with the rack angle α, i.e., $\beta = 90° - \alpha$ (Fig. 8.1b). Upon the design of louvers, Albayrak et al.

I. Albayrak (✉) · R. M. Boes
Laboratory of Hydraulics, Hydrology and Glaciology, ETH Zurich, Zurich, Switzerland
e-mail: albayrak@vaw.baug.ethz.ch

R. M. Boes
e-mail: boes@vaw.baug.ethz.ch

P. Rutschmann et al. (eds.), *Novel Developments for Sustainable Hydropower*,
https://doi.org/10.1007/978-3-030-99138-8_8

99

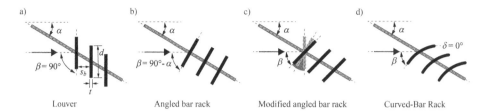

Fig. 8.1 Different fish guidance structure layouts with wide bar spacing **a** louver, **b** angled bar rack, **c** modified angled bar rack (MBR) and **d** curved-bar rack (CBR)

(2018, 2020) developed a Modified angled Bar Rack (MBR) with β independent of α, preferably $\beta = 45°$ instead of $90°$. Such a reduction of the bar angle reduces the head loss and improves the rack downstream flow field (Fig. 8.1c). Albayrak et al. (2020) reported the flow fields and fish guidance efficiencies (FGEs) of a louver with $\alpha = 15°$ and $s_b = 50$ mm and MBR configurations with $\alpha = 15°$ and $30°$, $s_b = 50$ mm and with and without bottom overlays for barbel (*Barbus barbus*), spirlin (*Alburnoides bipunctatus*), European grayling (*Thymallus thymallus*), European eel (*Anguilla anguilla*) and brown trout (*Salmo trutta*). The results show that MBR with $\alpha = 15°$ with and without overlay successfully guided 90% and 80% of the tested fish species, respectively. Furthermore, MBR with $\alpha = 30°$ with an overlay guided 95% of the tested fish. Such high FGEs and improved flow field of MBR led to the development of an innovative Curved-Bar Rack-Bypass System (CBR-BS) for a safe downstream fish passage at small- to large-scale HPPs and water intakes (Beck 2020; Beck et al. 2020a, b, c, Figs. 8.1 and 8.2).

8.2   Curved-Bar Rack-Bypass System

A CBR consists of vertical curved bars instead of straight bars used in louvers and MBR. They are arranged with equidistant spacing along the rack axis and mounted in a rack frame. The rack is placed across an intake canal at a rack angle typically $\alpha = 15$–$30°$ (Fig. 8.2a, b). A curved-bar is designed to have a bar angle to the flow direction ranging from $\beta = 45$–$90°$ at the upstream bar tip and an outflow angle of $\delta = 0°$, i.e. parallel to the flow direction in the power canal at the downstream end of the bar (Fig. 8.2d). The clear spacing between the bars is $s_b \geq 25$ mm, the bar thickness $t = 10$ mm, and the bar depth $d = 100$ mm. The upstream and downstream bar tips are typically rounded to avoid fish injuries.

A CBR creates hydrodynamic cues of turbulence, high velocity and pressure gradients by its bars similar to the working principle of louvers and MBR (Albayrak et al. 2020; Beck et al. 2020c). Such flow structures in front and between the bars are perceived and avoided by fish approaching the rack. Thanks to the angled rack arrangement, the velocity component parallel to the rack, V_p, guides the fish towards the bypass system

a)

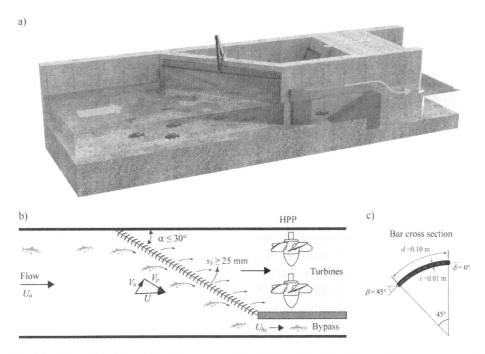

b)

c)

Fig. 8.2 Illustration (**a**) and detailed geometry (**b**) of Curved-Bar Rack-Bypass System and curved bar cross-section (**c**)

(BS) without causing a shock from a major physical contact at the rack. A CBR acts as behavioural barrier for smaller fish while it functions as a physical barrier for fish whose width is larger than the bar spacing (Fig. 8.2b). For an effective guidance of the CBR, the ratio between V_p and the rack normal velocity V_n should be above 1 along the rack, i.e. $V_n < V_p$ (Courret and Larinier 2008). Furthermore, to ensure that fish can swim actively along the CBR without exhaustion, the rack normal velocity should be smaller than the sustained swimming speed of fish, i.e. $V_n < V_\text{sustained}$. A general value of $V_\text{sustained} = 0.50$ m/s is recommended for smolts and silver eels (Raynal et al. 2013) as a first proxy.

Laboratory tests by Beck et al. (2020c) confirm the behavioural guiding effect of the CBR for several fish species except the European eel. They reported that above 75% of spirlin, barbel, nase (*Chondrostoma nasus*) and Atlantic salmon parr (*Salmo salar*) and below 75% of brown trout and eel were efficiently guided by a hydraulically optimized CBR configuration with $s_b = 50$ mm, $\alpha = 30°$ to a full depth BS in the laboratory tests. The use of bottom and top overlays may improve the FGE of the CBR-BS for bottom and surface-oriented fish species, respectively. The effectiveness of such overlays was demonstrated and recommended by EPRI and DML (2001) and Amaral (2003) for louvers and by Albayrak et al. (2020) for MBR. Furthermore, overlays can mitigate operational problems of driftwood, organic fine material and sediment by guiding them to the bypass (Beck 2020).

The curved-bars of a CBR cause a flow straightening effect, which results in ~20 and ~5 folds lower head losses compared to the same Louver and MBR configurations and in quasi-symmetrical downstream flow (Beck et al. 2020b), improving the rack downstream flow field and possibly HPP turbine efficiency. A head loss prediction equation for louvers, MBR and CBR is presented in Beck et al. (2020a).

Successful CBR design requires a good bypass system design, which should attract, safely collect and transport the fish and return them unharmed to the river downstream of a HPP. Full depth, surface, bottom and both surface and bottom bypasses are the main types and should be selected based on the biomechanical requirements of the target fish species and HPP layout. The ratio of the bypass entrance flow velocity to the approach flow velocity $VR = U_{by}/U_o$ and a gradual velocity increase along the rack to the bypass are crucial parameters for fish guidance and bypass acceptance (e.g. Simmons 2000; Albayrak et al. 2020; Beck 2020; Beck et al. 2020c). To this end, USBR (2006) recommends $1.1 \leq VR \leq 1.5$ for louver-BS, Ebel (2016) recommends $1.0 \leq VR \leq 2.0$ for horizontal bar rack-BS, while Beck et al. (2020c) recommend $VR = 1.1 \leq VR \leq 1.2$ for CBR-BS or other FGS to protect and guide fish of all species, life stages and sizes.

8.3   Conclusions and Outlook

Given the significantly reduced head losses and high fish guidance and protection efficiencies, CBR-BS presents a high potential over Louvers and MBRs for a safe downstream fish movement at HPPs at minimum negative economic impacts. Cost-effective engineering design recommendations for CBR-BS are given in-detail by Beck (2020). The first CBR-BS variant is currently installed, and its effectiveness will be assessed at the pilot HPP of Herrentöbeli located on River Thur in Switzerland. More projects at HPPs of different sizes and layouts are needed to evaluate the CBR-BS effectiveness under various flow conditions and for different fish species and to further improve its design.

References

Albayrak I, Kriewitz CR, Hager WH, Boes RM (2018) An experimental investigation on louvres and angled bar racks. J Hydraul Res 56(1):59–75. https://doi.org/10.1080/00221686.2017.1289265

Albayrak I, Boes RM, Kriewitz-Byun CR, Peter A, Tullis BP (2020) Fish guidance structures: new head loss formula, hydraulics and fish guidance efficiencies. J Ecohydraulics. https://doi.org/10.1080/24705357.2019.1677181

Amaral SV (2003) The use of angled bar racks and louvers for guiding fish at FERC-licensed projects. FERC fish passage workshop. Holden, USA

Bates DW, Vinsonhaler R (1957) Use of louvers for guiding fish. Trans American Fish Soc 86(1):38–57

Beck C (2020) Fish protection and fish guidance at water intakes using innovative curved-bar rack bypass systems. In: Boes RM (ed) VAW-Mitteilung, vol 257. VAW, ETH Zurich, Switzerland. https://vaw.ethz.ch/en/the-institute/publications/vaw-communications/2010-2019.html

Beck C, Albayrak I, Meister J, Boes RM (2020a) Hydraulic performance of fish guidance structures with curved bars: part 1: head loss assessment. J Hydraul Res 58(5):807–818. https://doi.org/10.1080/00221686.2019.1671515

Beck C, Albayrak I, Meister J, Boes RM (2020b) Hydraulic performance of fish guidance structures with curved bars: part 2: flow fields. J Hydraul Res 58(5):819–830. https://doi.org/10.1080/00221686.2019.1671516

Beck C, Albayrak I, Meister J, Peter A, Selz OM, Leuch C, Vetsch DF, Boes RM (2020c) Swimming behavior of downstream moving fish at innovative curved-bar rack bypass systems for fish protection at water intakes. Water 12(11):3244. https://doi.org/10.3390/w12113244

Courret D, Larinier M (2008) Guide pour la conception de prises d'eau 'ichtyocompatibles' pour les petites centrales hydroélectriques (Guide for the design of fish-friendly intakes for small hydropower plants). Agence de l'Environnement et de la Maîtrise de l'Energie (ADEME) (in French)

Ebel G (2016) Fischschutz und Fischabstieg an Wasserkraftanlagen – Handbuch Rechen-und Bypasssysteme. Ingenieurbiologische Grundlagen, Modellierung und Prognose, Bemessung und Gestaltung. In: Fish protection and downstream passage at hydro power stations – handbook of bar rack and bypass systems. Bioengineering principles, modelling and prediction, dimensioning and design), 2nd edn. Büro für Gewässerökologie und Fischereibiologie Dr. Ebel: Halle (Saale), Germany, 2016 (In German)

[EPRI] Electric Power Research Institute (US), [DML] Dominion Millstone Laboratories (US) (2001) Evaluation of angled bar racks and louvers for guiding fish at water intakes. Palo Alto (CA) and Waterford (CT): EPRI. Report No.: 1005193

Raynal S, Chatellier L, Courret D, Larinier M, Laurent D (2013) An experimental study on fish-friendly trashracks – part 2. Angled trashracks. J Hydraul Res 51(1):67–75

Simmons A (2000) Effectiveness of a fish bypass with an angled bar rack at passing Atlantic salmon and steelhead trout smolts at the Lower Saranac Hydroelectric Project, advances in fish passage technology. Am Fish Soc 95–102

USBR (2006) Fish protection at water diversions – a guide for planning and designing fish exclusion facilities. Technical Report. U.S. Department of the Interior, Bureau of Reclamation

Guidelines for Application of Different Analysis Methods of Fish Passage Through Turbines—Impact Assessment of Fish Behavioural Aspects

9

Franz Geiger and Ulli Stoltz

9.1   Introduction

The fast-moving turbine parts create challenging hydrodynamic conditions for a fish passing through a turbine. Hence, several modelling approaches have been developed to assess the impact on the fish during a turbine passage, which potentially causes injuries. Survival rates can be predicted based on the fish species, the turbine main dimensions and the operating condition.

The most simplified modelling approach is statistical modelling of a set of experimental data e.g. (Larinier and Travade 2002). Such empirical models are based on specific field test data. Therefore, transferability to other application cases is only limited. To provide a more general applicability, a number of models have been developed throughout the decades, which account for the physical conditions and the resulting biological response of the fish. Such models are available for the most common turbine types, i.e. Kaplan and Francis turbines.

Collisions of the fish with the rotating runner blades, so-called strike events, are the most evident source of damage. Thus, most models evaluate the risk of exposure to relevant physical load by calculating the runner blade strike probability. This can be derived by theoretical considerations of a simplified geometric object that passes through the runner blades. Such strike probabilities can facilitate basic assessment of overall damage

F. Geiger
Ecohydraulic Consulting Geiger, Wallgau, Germany
e-mail: franz.geiger@hycor.de

U. Stoltz (✉)
Hydraulic Development, Voith Hydro Holding GmbH & Co. KG, Heidenheim, Germany
e-mail: ulli.stoltz@voith.com

P. Rutschmann et al. (eds.), *Novel Developments for Sustainable Hydropower*,
https://doi.org/10.1007/978-3-030-99138-8_9

rates. Different models have been developed for blade strike probability calculation. Most eminent are those of von Raben (1958) and Montén (1985).

In recent years more enhanced methods based on numerical simulations (e.g. Richmond et al. 2014) were developed allowing the analysis of typical physical impact variables as stress, shear and barotrauma of the downstream fish passage. These physical impacts can then be correlated to the biological impact on the fish.

In the following, enhancements of both simple und advanced modelling methods are investigated. Additionally behavioural aspects are considered in the modelling as these influence the risk of potential injuries in a passage event. The modelling methods are applied to the FIThydro Testcases in Bannwil, Guma and Obernach, and analyzed at representative operating conditions.

9.2 Introduction to Testcases

The impact of turbine passage on fish depends not only on the turbine type, which is applied at a certain power plant. Even for the same turbine type, several factors, like size, rotational speed, number of runner blades and operation mode, play an important role. To accurately judge the effect of these factors on turbine fish passage a more detailed evaluation of each particular situation is required. In the following, the focus is on typical run-of river power plants, which mostly are equipped with axial turbines, either with movable blades or as propeller machine. The investigation performed in the FIThydro project focusses on bulb turbines. In Table 9.1 a brief overview on the main parameters of the Testcase power plants is given.

In the following, the modelling approaches are briefly introduced. Then some exemplary results of the research at the FIThydro sites are presented. More detailed results of both the fish passage modelling and the experimental tests are presented in the FIThydro documentation (Geiger and Stoltz 2019).

Table 9.1 Main parameter of testcase turbines

Power plant	Bannwil	Guma	Obernach
Power output (per unit) (MW)	9.5	1.8	0.035
Rated discharge (m^3/s)	142	25	1.5
Rated head (m)	7.2	8.0	2.5
Runner diameter (m)	4.4	2.1	0.75
Runner speed (rpm)	107	220	333
Number of blades	4	4	4
Turbine type	Bulb	S-turbine	Bulb

9.3   Simple Modelling Methods

Based on the existing simple modelling approaches for strike probabilities, recent work on generalized modelling (Geiger 2018) showed, that model differences originate from different assumptions for fish body alignment to the flow direction close to the turbine. This aspect is related to the fish behaviour during turbine passage, which is still not sufficiently understood. For typical turbine designs and operating conditions, the different blade strike models provide comparable results. The comparison of the simple modelling approaches with CFD based analysis showed that the modelling accuracy depends strongly on hydraulic conditions. Opening angles of runner blades and corresponding flow angle close to the runner blade entrance edge need to be estimated correctly. These values are frequently not directly available. Therefore, several authors provided estimates or empirical formulas. As the design of turbines is based on fundamental laws of physics, design guidelines for hydraulic machines can provide well-estimated values for blade strike modelling. This allows an impact assessment with simple and fast methods respecting the underlying physical relations and engineering procedures. The comparison of this method with 3D-CFD results for the FIThydro Testcases showed a better agreement for parameters like the flow angles. Therefore, it provides more accurate modelling results and a broader applicability, compared to previous empirical models (Geiger et al. 2020a).

In a second step survival rates can be derived based on the runner blade strike probabilities. As it is well known that not every collision of fish with a runner blade results in relevant injury further aspects need to be taken into account. If the relative speed of the fish in relation to the runner is small, the impact of a strike event is less significant. Thus, the biological response of a fish species is playing a role. In the 20th century, these relations were included in the modelling approaches by statistical methods using empirical data of test results at specific hydropower plants (e.g. von Raben (1958), Montén (1985)). However, this approach causes transferability problems, especially for untypical turbine setups or operating conditions.

In the last decades, experimental campaigns were conducted to gain detailed reference information on resulting fish injury in case of physical load, for example in function of blade shape and blade strike velocity (Turnpenny et al. 2000; Amaral et al. 2011). This biological response data can be combined with existing blade strike models to improve the modelling method. As blade strike related mechanical injury is usually the dominant impact source for fish passage at low head run-of-river HPP, this simplification can provide useful information.

The basic relations between hydraulic boundary conditions, runner blade strike probabilities and biological response enable a simple assessment of fish damage rates during turbine passage as shown in Fig. 9.1. Comparison with literature references shows typical accordance of predicted mortality rates within a magnitude of about 10 % of the experimentally observed values (Geiger et al. 2020a). The unresolved accuracy of experimental

Fig. 9.1 Flow chart of the main steps and aspects of damage rate modelling

data remains an issue in this context, as well as the influence of the actual fish behaviour during turbine passage.

The most common assumption for modelling purpose is that a fish passes through the turbine at mid-radius; the fish body is aligned with the flow and has the identical velocity, without active swimming speed. A comparison of model results with experimental observation of damage rates suggests that this assumption provides typically suitable accordance. The generalized modelling also allows for correct modelling of different passage locations, orientations and speed.

The simple modelling methods enable a fast and inexpensive quantification of damage rate magnitudes. They do not require special hardware, software or detailed information, for example about the turbine geometry. Only basic turbine and power plant parameters like runner diameter, number of runner blades, rotational speed, discharge and head are required. In order to improve fish passage by tailor-made runner blade designs for enhanced fish passage, simple modelling approaches are not advisable. The exposure to hydrodynamic forces and the impact on the fish during the turbine passage can only be roughly estimated. The comparison within the FIThydro project also showed systematic deviations of such simplified modelling approaches from the detailed 3D CFD evaluations (Geiger and Stoltz 2019).

9.4   Advanced Modelling Methods Using CFD

To capture detailed information on hydrodynamic conditions during turbine passage the numerical modelling is most accurate. A widely used approach is to perform steady state CFD modelling and to analyse the simulation results following streamlines and extracting the physical relevant information concerning strike, pressure, turbulence and shear along them. Besides this rather simple procedure, other approaches using transient CFD are also available, which use i.e. particle tracks to evaluate the strike risk. As these models are rather complex and time-consuming to apply, it is not suitable as standard procedure during the design process of a water turbine. Hence, the focus of the research work in the FIThydro project was to identify a process suitable for industrial applications.

The evaluation of the fish passage assumes that the pathway of a fish follows a streamline through the turbine. These streamlines are generated with a stationary CFD simulation and are then post-processed with a tool applying the biological performance assessment (BioPA) developed by Pacific Northwest National Laboratory (PNNL) (Richmond et al.

2014). The information for the stressor exposure of pressure, strike, shear and turbulence can be extracted directly from the streamlines. For strike, the velocity vectors close to the blade entrance edge are used in order to calculate the strike probability and the impact velocity a fish experiences when colliding with the blade. For the other stressors an exposure probability is derived based on a large number of streamlines. As presented in Fig. 9.2, the injury risk can then be derived by combining the physical information with dose response data of respective fish species. A score is then integrated over the product of exposure probability and exposure mortality of the fish. The value is high when the risk of passage injury is low. It is understood that for now, the score does not represent an absolute passage-survival estimate. However, it offers a systematic way to evaluate trade-offs associated with various hydraulic solutions. An optimization of the hydraulic shape of the turbine can then be performed in order to reduce the risk of an injury during turbine passage (DeBolt et al. 2015).

As a part of the FIThydro project, investigations at the laboratory Testcase in Obernach were performed. The impact of turbine passage on brown trout of various fish lengths was investigated for the Kaplan Bulb-Turbine at different operating modes (Geiger et al. 2020b).

Figure 9.3 presents the results of the analysed full load point, including a variation

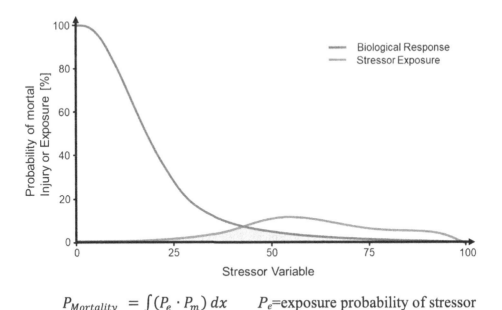

$$P_{Mortality} = \int (P_e \cdot P_m)\, dx \qquad P_e = \text{exposure probability of stressor}$$
$$P_m = \text{exposure mortality of fish}$$

Fig. 9.2 Example of BioPA evaluation with probability distribution and biological dose-response of stressor variable as typically applied (see also Richmond et al. 2014)

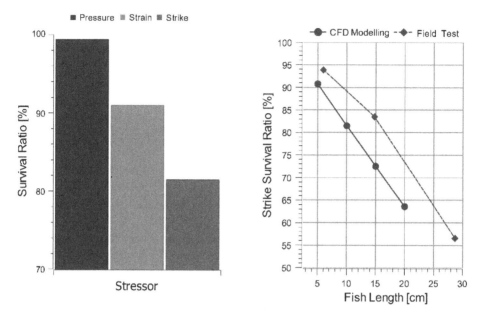

Fig. 9.3 Results of CFD modelling in comparison to field test for the rated operating point with $H = 2.5$ m and $Q = 1.5$ m^3/s at the Obernach Testcase

of fish length. The results underline the importance of reasonable definition of the relevant fish length, depending on life stages of fish, fish species or trash rack clearances. The field test assessment of the fish passage by the team of the Technical University of Munich could be confirmed by the CFD based evaluation. The identified injuries at site were mainly caused by strike events, no clear indication for barotrauma and shear related injuries could be found. This was also confirmed by the CFD-based modelling method; only the shear stress influence seems to be over predicted. As presented in Fig. 9.3 the strike modelling also exceeds the values of the experimental results. Eye catching is the fact that a continuous shift in the results seems to be present.

Looking into existing studies we came across the results of experimental tests in the Oak Ridge Laboratory (Bevelhimer et al. 2017), which indicate that tail strikes rarely lead to an injury of the fish. This corresponds well to the principal definition of fish body region of a trout in which the tail region is approximately 1/3 of the body. Accordingly adapting the effective fish length by a factor of 2/3, improves the agreement of the strike modelling to the experiments significantly and leads to an excellent match of the datasets. Additionally, other effects like fish orientation also partly contribute to this effect.

This leads to another special focus within the FIThydro project, the modelling of fish behavioural effects. The research goal was to evaluate the significance of these influences on the impact a fish experiences during turbine passage.

These effects could be studied for the test site Bannwil in Switzerland and the Guma power plant in Spain. In a first phase, tests with BDS sensors were performed for both power plants. The sensors were injected at the intake of the power plant and passed neutrally buoyant with the flow through the turbine recording pressure data. This data was then compared with a standard BioPA analysis of the same operating conditions. The results showed a good agreement for the assessment of the nadir pressures. In the second phase, the modelling was extended to cover the effect of behavioural aspects.

One interesting consideration is the passage location of a fish. Three passage locations as illustrated in Fig. 9.4 at inside, middle and outside having the same area were used to analyse several operating conditions.

The results in Fig. 9.5 show a representative operating point for the Bannwil power plant. It can be seen that the passage location influences the survival ratio based on the different stressors. The results show that the passage location has a significant influence on survival. A passage in the middle of the blade is favourable regarding pressure and

Fig. 9.4 Variation of passage location coloured by strike survival ratio from low (blue) to high (red)

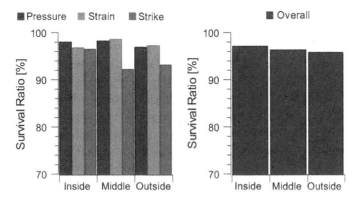

Fig. 9.5 Variation of passage location at an operating point close to the optimum at the Bannwil testcase

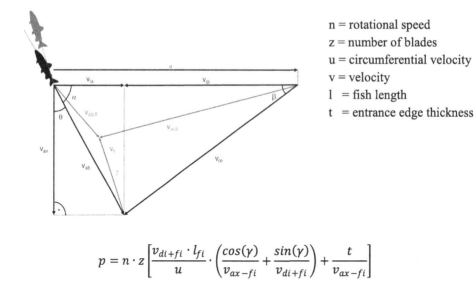

$$p = n \cdot z \left[\frac{v_{di+fi} \cdot l_{fi}}{u} \cdot \left(\frac{cos(\gamma)}{v_{ax-fi}} + \frac{sin(\gamma)}{v_{di+fi}} \right) + \frac{t}{v_{ax-fi}} \right]$$

$$v_{ax-fi} = v_{ax} - v_{fi} \cdot cos(\gamma)$$

$$v_{di+fi} = v_{di} + v_{fi} \cdot sin(\gamma)$$

Fig. 9.6 Velocity triangle and strike formula including fish velocity and orientation (Geiger 2018)

shear influences and a passage close to the hub is favourable especially regarding strike, due to the low impact velocities.

Besides the passage location, also the fish orientation and potential swimming speed influence the strike rate. Based on the assumption that a fish swims against the main flow direction and maintains his swimming depth the basic strike formula is extended as presented in Fig. 9.6 (Geiger 2018).

Typical flow velocities during turbine passage are low in the intake region, but as soon as the flow approaches the guide vanes the flow accelerates and velocities increase rapidly above more than 2 m/s, which is typically the velocity most fish can swim against for a certain amount of time. Faster velocities are only possible for short sprints. Like this most fish species will barely be able to withstand the flow conditions and have only limited capabilities to control or influence turbine passage trajectories in close vicinity of the turbine. Nevertheless, a principal study is performed, and the results presented in Fig. 9.7 indicate that a fish actively swimming against the main flow increases the passage time through the turbine and therefore the strike probability. At the same time the impact velocities increase slightly. Therefore, the overall risk of a strike injury rises.

Regarding the orientation of the fish in relation to the blade a derivation of the von Raben correlation (von Raben 1958) is used, which assumes that the fish is oriented with the flow. Variations of the orientation as applied in Fig. 9.7 are presented in relation to

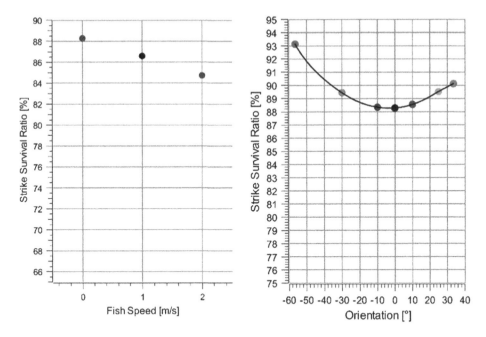

Fig. 9.7 Influence of fish swimming speed and fish orientation on survival rate at full load operation at the Guma power plant

the absolute flow direction, which is indicated with 0°. A range from a radial to an axial orientation is considered, whereas an angular deviation of more than 45° to the main flow is highly unlikely. Analysing the results, the standard angle of 0° is a rather conservative approach, with a maximum tolerance of 1–2% expected for the fish passage assessment. Benchmarking the different influencing factors, fish length and passage location have a much higher impact as the orientation on the survival rate modelling. In general, the importance of any impact also depends on turbine size and speed, as well as the operating conditions. Site-specific considerations are advisable.

9.5 Summary and Outlook

During the FIThydro project simple and more complex modelling methods were applied. Depending on the purpose and the stage of the development of a power plant, the right method needs to be chosen. Especially in early stages of a project simple methods are sufficient to gain a general overview, however it is recommended to include influencing factors in the analyses to avoid a blurry picture of the situation.

It is important to represent the hydraulic boundaries as accurate as possible by applying relevant operating conditions. In addition, relevant fish species need to be identified to judge the biological sensitivity correctly and to set the right focus for mitigation measures. More advanced CFD based methods enable an enhanced turbine design to improve fish passage conditions significantly. Projects like the Ice Harbor power plant at the Snake River (Foust et al. 2013) in the US show that it is already possible to apply these modelling methods successfully in the turbine design process.

In the future, a good and accurate modelling in combination with the application of sensors might avoid the need of life fish tests. An Assessment of possible impacts on the ecology of river reaches is already feasible in early stages of a hydropower project. Accordingly, if needed mitigation measures to minimize the influences can be designed.

Acknowledgements Special thanks to Carl Robert Kriewitz and Manuel Henzi of BKW (Bannwil) and Juan Carlos Romeral de la Puente (Guma) of SAVASA for the support with the sensor fish tests as well as making the geometry and operational data available for the numerical modelling. Also we would like to thank Geppert GmbH for providing the hydraulic geometry of the Obernach turbine for the numerical modelling.

References

Amaral S, Hecker G, St. Jean S, McMahon B (2011) 2010 Tests examining survival of fish struck by turbine blades. Tech Rep. EPRI, Palo Alto

Bevelhimer MS, Prachell BM, Fortner AM, Deck KL (2017) An overview of experimental efforts to understand the mechanisms of fish injury and mortality caused by hydropower turbine blade strike. Oak Ridge National Laboratory

DeBolt D, Donelson DK, Richmond M, Strickler B, Weisbeck M (2015) Development of Priest Rapids turbine upgrade project. Hydrovision, Portland, USA

Foust JM, Donelson RK, Ahmann M, Davidson R, Kiel J, Freeman T (2013) Improving fish passage on the snake river: turbine development at Ice Harbor lock and dam. Proceedings HydroVision Conference Sacramento, USA

Geiger F, Stoltz U (2019) D3.1—Guidelines for mortality modelling. FIThydro Project Report. https://www.fithydro.eu/deliverables-tech/

Geiger F, Cuchet M, Rutschmann P (2020a) Zur Berechnung der Schädigungsraten von Fischen bei der Turbinenpassage. Wasserwirtschaft, Dec 2020a

Geiger F, Cuchet M, Rutschmann P (2020b) Zur Verringerung der Schädigungsraten von Fischen bei der Turbinenpassage. Wasserwirtschaft, Dec 2020b

Geiger F (2018) Fish mortality rate during turbine passage—generalized runner blade strike probability modelling. 12th International Symposium on Ecohydraulics, Tokyo, Japan

Larinier M, Travade F (2002) Downstream migration: problems and facilities. In: Bulletin Français de la Pêche et de la Pisciculture, 364 supplément, pp 181–207

Montén E (1985) Fish and turbines: fish injuries during passage through power station turbines. Stockholm, Vattenfall

Richmond M, Serkowski J, Rakowski C, Strickler B, Weisbeck M, Dotson C (2014) Computational tools to assess turbine biological performance. Hydro Review 33:88

Turnpenny AW, Clough S, Hanson KP, Ramsay R, McEwan D (2000) Risk assessment for fish passage through small, low-head turbines. Tech. Rep. United Kingdom

Von Raben K (1958) Zur Beurteilung der Schädlichkeit der Turbine für Fische. Wasserwirtschaft 48:60–63

Measures to Improve Fish Passage Through a Turbine

Franz Geiger, Peter Rutschmann⊙, and Ulli Stoltz

10.1 Introduction

Besides the economic performance, the ecologic footprint of a turbine is becoming more and more relevant. One impact of hydropower usage on river ecology is related to fish damage during turbine passage (c.f. Chap. 9). Therefore, several mitigation measures have been developed to avoid fish passage through turbines, e.g. by screening and bypass systems (c.f. Chaps. 7 and 8). Moreover, special turbine types and designs can provide improved passage conditions for fish (e.g. Chap. 9). These approaches are often related to high construction costs and maintenance efforts as well as drawbacks in energy production. For numerous sites, the technical feasibility is questionable, especially regarding various existing hydropower plants and within suitable time scales.

Therefore, alternative approaches are desirable, which enable cost and energy efficient ecological improvement, and which can be implemented in short time scale. Instead of avoiding fish passage through the turbine or replacing the existing turbine, potentially combined with a powerhouse upgrade, the strategy of such approaches should target to improve the interaction of turbine and fish. Modelling fish damage probabilities during

F. Geiger (✉)
Ecohydraulic Consulting Geiger, Wallgau, Germany
e-mail: franz.geiger@hycor.de

P. Rutschmann
Hydraulic and Water Resources Engineering, Technical University of Munich, Munich, Germany
e-mail: peter.rutschmann@tum.de

U. Stoltz
Voith Hydro Holding GmbH & Co. KG, Hydraulic Development, Heidenheim, Germany
e-mail: ulli.stoltz@voith.com

© The Author(s) 2022
P. Rutschmann et al. (eds.), *Novel Developments for Sustainable Hydropower*,
https://doi.org/10.1007/978-3-030-99138-8_10

turbine passage (see Chap. 9) provides a detailed understanding of damage risk relations and offers opportunities to reduce these risks and the associated ecological impacts.

On the one hand, this goal can be addressed by adapting the turbine operating conditions. The physical impact on fish can be reduced by changing the hydraulic conditions during the turbine passage and with adapted blade and guide vane openings also mechanical risk can be reduced. On the other hand, the behaviour of the fish influences the damage rates during turbine passage, as they depend on the passage location, body orientation and potential swimming speed of the fish. Therefore, influencing the fish behaviour prior or during the passage process can also reduce damage risks.

Depending on the hydropower site, such techniques potentially provide a feasible alternative for the above-mentioned measures or they can be combined with the classical approaches to further improve fish passage conditions at sensitive sites, e.g. regarding small fish which cannot be addressed by screening systems.

10.2 Changing Operation Modes for Reduced Risk—Best Practice Guidelines for Turbine Operation

Fish passage through the turbine is not only dependent on factors like machine size, number of blades and rotational speed but also on the influence of different operating conditions. Therefore, it is beneficial to assess the impact of changing operating conditions on the different physical stressors acting on the fish during turbine passage. The generation of so-called fish passage hill-charts provides the potential to adapt operation based on migration periods of different fish species.

In order to identify the individual damage mechanisms, the CFD based fish passage modelling as presented in Chap. 9 was applied. Besides the localization of critical passage regions, the evaluation of a complete set of operating points helps to better identify which parameters affect the fish passage at a certain operating scheme. As part of the research within the FIThydro project, a representative set of operating points were analysed for the Testcases Bannwil and Guma. This range was extended from the original operating range in order to obtain a significant influence of each stressor and to identify trends. The focus was to determine the influence of the stressors such as nadir pressure, strike, shear and turbulence on fish. For a better representation of the results, the BioPA performance score was converted into a scoring system from 0 to 10, where 10 is the best score and corresponds to the BioPA score of 100. The scoring of 0 corresponds to 70 in BioPA.

The performed study used a representative fish length of 10 cm and the biological response data for salmon. For both Testcases, we used the boundary conditions based on operational data of recent years. Special care was necessary to use the correct water levels and acclimation depth of the fish for each operating condition to represent a realistic pressure level, when analysing barotrauma. For example, for the Bannwil HPP, the pressure score was calculated based on an acclimation depth of 5 m, the tail water level was calculated based on a constant head water level and the head of the respective operating point.

The calculated survival rates for turbulence based on the biological response of salmon were close to 100%. As no influence of operating conditions was apparent, the results were not analysed further. However, secondary effects like disorientation are not in the scope of this study.

Figure 10.1 presents exemplarily the results of the Bannwil turbine, showing the hill-charts of the different stressors, as well as a combined score with equally weighted factors. It is obvious that the influence of the different stressors on fish survival is varying with the operating condition. Large blade openings with high discharge show a reduced risk related to strike. The effect of nadir pressure is oppositional as for large flows low-pressure zones especially close to the runner blade are present and the pressure change during the turbine passage increases. Strain is closely related to flow separation zones and bad flow quality. This is not only dependent on the hydraulic shape of a turbine, but also on the operating condition as seen in Fig. 10.1. The results of the Guma Testcase show in principle the same tendency as the analysis of the Bannwil machine (Geiger and Stoltz 2019). Depending on different boundary conditions like size, rotational speed and machine type, as well as fish length and acclimation depth the hill charts differ in value and peak.

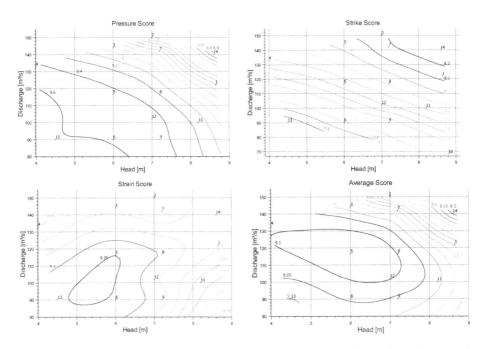

Fig. 10.1 Bannwil HPP—Individual fish passage hill-charts and combined chart using equally weighted stressors

In summary, one can say that the fish passage hill-charts can be helpful to adapt the power plant operation while a migration period of a certain fish species to improve the fish passage through the turbine. Additionally, the stressor variables are judged individually to adapt the operational scheme of the power plant as function of the individual susceptibility of the relevant species. Finally focusing on the most significant operating conditions improves the hydraulic design. All relevant impacts while operation can be taken into account for the design of an optimized turbine blade for an enhanced fish passage through the turbine.

10.3   Influencing the Impact on Fish During Turbine Passage—IDA

The work presented in Chap. 9 showed that the survival ratio of fish during turbine passage depends on the passage position, fish orientation and swimming speed. The actual impact of the different aspects depends on the site-specific conditions. For example, Fig. 10.2 shows on the left the damage probability dependent on passage radius and discharge for a 2 MW Kaplan turbine. Experimental fish passage investigations e.g. by Geiger et al. (2020a), indicate that fish naturally do not pass the turbine in a favourable way. Based on such findings and further considerations, an innovative method was developed and patented at TUM (EU Patent EP3029203): The fish protection system IDA (Induced Drift Application) increases the survival probability of fish during a turbine passage by influencing the fish behaviour.

The behaviour of fish can in principle be influenced by stimuli like light, sound or electric fields as suggested e.g. by Kruitwagen 2014, Jacobson et al. 2017, Murchy et al. 2017, Sonny et al. 2006 or Parasiewicz et al. 2016. Electric fields are especially favourable for IDA implementation. In general, low electric field strengths exert a repulsive effect on

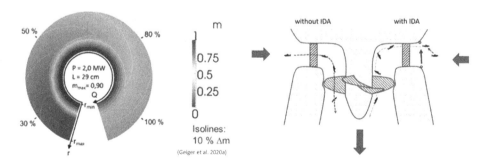

Fig. 10.2 Modelled damage rates m for an exemplary 2 MW Kaplan turbine, in dependency of passage radius r, discharge Q (normalized by design discharge) and fish length L (total length); m_{max} provides the maximum value observed (left) and schematic presentation of fish passage through a vertical axis Kaplan turbine without and with IDA implementation (right) (see Geiger et al. 2020a)

fish. In contrast, fish move towards the anode at high field strengths, so-called galvano-taxis. Depending on the strength of the electric field and period of time the fish are exposed to it, an electric field can cause a shorter or longer anesthesia of a few seconds to a few minutes. The reaction of fish is already technically used for behavioural barriers, electro-fishing and electro-narcosis. When applied correctly, these effects do not harm the fish.

Therefore, an adequate electric field can achieve exactly a desirable effect in terms of an increased fish survival probability: Fish are directed against the turbine hub where the survival probability is higher if the anode is placed there. In addition, fish are briefly impaired in their ability to swim by electro-narcosis, which means that they are exposed to the strike probability risk for less time and that their body is randomly angled to the streamlines, which results in higher survival probabilities. Moreover, the field strengths should be minimized as much as possible to avoid an unintentional exposure of the fish to damage by predators after turbine passage.

Figure 10.2 shows on the right the schematically depicted difference without and with the principle of galvano-taxis exerted to fish. In the course of the FIThydro project, the IDA invention was tested for the first time with live fish experiments on a small Lab prototype using a Bulb turbine at the Lab of TUM. The fish behaviour was influenced with an electric field for these investigations and yielded promising results. The exemplary considerations were based on strike damage considerations for Kaplan turbines, which is the most relevant aspect for run-of-river power plants with rather small heads like the FIThydro Testcases. The IDA principle can be adapted to other turbine types and the positioning and field strengths can be optimized on a case-by-case basis. Also, it can be combined with other stimuli, like light and sound.

The first results with IDA were obtained by carrying out tests on a 35 kW Kaplan Bulb turbine in the hydraulic engineering laboratory of TUM in Obernach. The experiments were conducted in the scope of an animal experiment permit (ROB-55.2-2532.Vet_02-19-66). 1201 Brown trout (*Salmo trutta fario*) with a total length from 5 to 30 cm were deployed. Two series of experiments were conducted with and without IDA. After passing through the turbine, all fish were collected, and injuries and mortalities were recorded. To also record non-visible damage due to internal injuries, all fish were observed for 96 h. The mortality or injury rates include those fish with relevant injuries, which put in question the long-term well-being of the fish. This corresponds to all categories except 1A and 2A in the classification according to the working aid of the German "Forum Fischschutz und Fischabstieg" (Schmalz et al. 2015).

The generation of an electric field required two electrodes and a power/voltage supply unit, with which the existing Kaplan turbine was retrofitted. Figure 10.3 shows the two ring-shaped copper electrodes at the inlet to the turbine. Their arrangement was tailored for the respective turbine type. In the case of the Bulb turbine in the TUM Lab, the ring-shaped anode was mounted around the Kaplan turbine hub to direct the fish to the passing location where their probability of survival is highest.

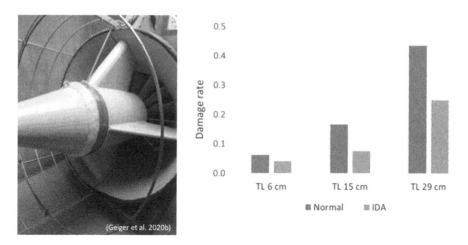

Fig. 10.3 The electrode setup at the bulb turbine intake (left) and exemplary results for damage rates (right) without IDA (normal) and with IDA (IDA) in dependency of the fish total length (TL) (see Geiger et al. 2020b)

The first results of IDA studies are provided in Fig. 10.3 (right). They show a reduction of the fish damage rate to about 55% and for all fish size classes compared to turbine operation without IDA. While the IDA efficiency was found to be limited by spatial aspects in very small-scale turbines, even higher ecological improvements can be expected for turbines of larger size. Further improvement can be achieved by optimizing the electric field strength and the combined use of other behavioural stimuli.

The IDA technology provides an alternative approach to reduce fish damage rates during turbine passage in order to improve ecological conditions of fish populations. In this context, it has to be acknowledged, that solutions for fish downstream passage at HPPs in general cannot and do not intend to fully protect downstream passing fish as this is not feasible. While damage to individual fish is justifiable given the benefits of hydropower production, damage rates have to be restricted to enable sustainable fish populations. Accordingly, the IDA technique reduces damage quota. It should be noted, that even mechanical barriers can only address fish of respective size, while smaller fish are subjected to turbine passage and corresponding damage.

The IDA's potential of damage rate reduction is turbine specific as the damage rates themselves. For given HPP sites, the allowable damage rates, the common fish damage rates for the particular turbines and operating conditions and the IDA damage reduction need to be assessed individually. For suitable cases, the IDA technology can provide a reduction of turbine related damage rates from non-allowable to acceptable values for sustainable fish populations.

At the same time, the IDA technology is associated with low construction, maintenance and servicing costs, especially compared to conventional screen and bypass systems. It has the further advantage that it does not affect power generation and can be easily and cost-effectively retrofitted in existing hydropower plants of medium and large capacity. The use of the IDA technology requires consideration of national animal welfare laws and appropriate authorization for its use, as well as consideration of intellectual property rights.

Moreover, the IDA technology can also be combined with conventional mitigation strategies e.g. be employed to reduce damage rates of small fish, which can pass through mechanical barriers and enable larger bar clearance for mechanical barriers. Further research and development are recommended. The achieved results provide promising perspective and show further potential for improvements. The installation and investigations of a prototype facility of larger size is advisable.

Acknowledgements Special thanks to Carl Robert Kriewitz and Manuel Henzi of BKW (Bannwil) and Juan Carlos Romeral de la Puente (Guma) of SAVASA for the support with the sensor fish tests as well as making the geometry and operational data available for the numerical modelling. Also, we would like to thank Geppert GmbH for providing the hydraulic geometry of the Obernach turbine for the numerical modelling.

References

Geiger F, Stoltz U (2019) D3.1—Guidelines for mortality modelling. FIThydro project report. https://www.fithydro.eu/deliverables-tech/

Geiger F, Cuchet M, Rutschmann P (2020a) Zur Berechnung der Schädigungsraten von Fischen bei der Turbinenpassage. Wasserwirtschaft Dec 2020a

Geiger F, Cuchet M, Rutschmann P (2020b) Zur Verringerung von Fischschäden in Turbinen mittels Verhaltensbeeinflussung. Wasserwirtschaft Dec 2020b

Jacobson P et al. (2017) Recent research on the effect of light on outmigrating eels and recent advancements in lighting technology. Tech. rep. Electric Power Research Institute (EPRI)

Kruitwagen G (2014) Research at IJmuiden lock complex provides unique insight in fish guidance. WIT Transactions on State-of-the-art in Science and Engineering 71

Murchy KA, Cupp AR, Amberg JJ, Vetter BJ, Fredricks KT, Gaikowski MP, Mensinger AF (2017) Potential implications of acoustic stimuli as a non-physical barrier to silver carp and bighead carp. Fish Manage Ecol 24:208–216

Parasiewicz P, Wiśniewolski W, Mokwa M, Zioła S, Prus P, Godlewska M (2016) A low-voltage electric fish guidance system—NEPTUN. Fish Res 181:25–33

Schmalz W, Wagner F, Sonny D (2015) Arbeitshilfe zur standörtlichen Evaluierung des Fischschutzes und Fischabstieges. Forum, Fisch Und Fisch

Sonny D, Knudsen FR, Enger PS, Kvernstuen T, Sand O (2006) Reactions of cyprinids to infrasound in a lake and at the cooling water inlet of a nuclear power plant. J Fish Biol 69:735–748. https://doi.org/10.1111/j.1095-8649.2006.01146.x

Archimedes Screw—An Alternative for Safe Migration Through Turbines?

Ine S. Pauwels, Jeffrey Tuhtan, Johan Coeck, David Buysse, and Raf Baeyens

11.1 Introduction

The Archimedes pump is one of the oldest feats of engineering still being used today. In recent times, it has seen a major revival in modern engineering, by reversing it for use as turbine (Waters and Aggidis 2015).

Archimedes turbines are frequently considered more "fishfriendly" than conventional turbines due to their very low rotational rates (30 rpm) and blade tip speeds (3.8 m/s), low rates of pressure change, low fluid shear, and a low overall number of blades reducing contact probability. But this considered fish-friendliness has only been examined in a handful of studies. Hence, many unanswered questions on the fish friendliness of Archimedes turbines remain. For instance, it is unknown if and how the harmfulness of the screws depends on the operation of the screw (e.g. do the rates of injury and mortality decrease if we operate the screw at a low rotational speed over a longer period of time)? Besides, it

I. S. Pauwels (✉) · J. Coeck · D. Buysse · R. Baeyens
Research Institute for Nature and Forest (INBO), Brussels, Belgium
e-mail: ine.pauwels@inbo.be

J. Coeck
e-mail: johan.coeck@inbo.be

D. Buysse
e-mail: david.buysse@inbo.be

R. Baeyens
e-mail: raf.baeyens@merck.com

J. Tuhtan
Department of Computer Systems, Tallinn University of Technology, Tallinn, Estland
e-mail: jetuht@ttu.ee

© The Author(s) 2022
P. Rutschmann et al. (eds.), *Novel Developments for Sustainable Hydropower*,
https://doi.org/10.1007/978-3-030-99138-8_11

is not clear how the characteristics of the screw influence the rates of injury and mortality, and if the potential relations differ per species. It may be the case that smaller screws pose an increased risk of injury and mortality than larger screws, and in general, screws may be less injurious when installed with a lower angle of inclination.

There are multiple ways to investigate the fish-friendliness of Archimedes screws at site. Until present scientists have been examining this by observations of injury and mortality of life fish (Schmalz 2010), or by sensors that sense the hydraulic forces fish are exposed to during passage (Boys et al. 2018). A few studies have combined life fish and sensor experiments (Deng et al. 2005), but not on an Archimedes turbine yet (Pauwels et al. 2020). So, the relation between the results of life fish studies and sensor studies also remains to be conclusively investigated.

Whether at new hydropower projects, or at sites where old turbines reach the end of their life and require refurbishing or replacement, there is considerable opportunity to further develop and optimize technologies and drive better outcomes for fish passage (Boys et al. 2018). Therefore, governments, policy makers, river managers and turbine designers need a list of the causes to design, build and remediate screws, to ensure that they provide a truly fishfriendly installation at each site. This requires much more multi-species analyses, including sensor analyses of multiple Archimedes screws of different dimensions and operational modes.

Within the FIThydro project, we investigated the rates of injury and mortality by multi-species fish experiments and the physical forces by barotrauma sensors during downstream passage through a large Archimedes hydrodynamic screw (10 m head, 22 m length and 3 m width, 1 MW). It was the first study to investigate multiple species, to combine life fish and passive sensor data and to investigate this in such a large Archimedes hydrodynamic screw.

11.2 Fish Passage at Archimedes Screws

Archimedes screws are among the world's oldest hydraulic machines that are still used today. Their primary use is as a type of low elevation water pump. In the latter part of the twentieth century, the screw re-emerged as a turbine (Waters and Aggidis 2015). In 1994, the first Archimedes screw turbine was installed in Europe, and by 2012 Lashofer et al. counted some 400 worldwide (Lashofer et al. 2012). Archimedes screw turbines are classified as small (1–10 MW) or mini (<1 MW) hydropower plants and are typically used at sites with a total elevation difference of 8–10 m and for discharges of 1–10 m^3/s (Quaranta and Revelli 2018). The screws rotate around an inclined axis ranging from $22°$ to $35°$ from the horizontal. They are further classified as "hydrodynamic screws" when the external cover does not turn with the screw, but is fixed and acts only as a support (Waters and Aggidis 2015; Quaranta and Revelli 2018; Lubitz et al. 2014) see Figs. 11.1 and 11.2.

Fig. 11.1 The hydropower station of Ham (Belgium), equipped with three Archimedes hydrodynamic screws (on the left; the cover does not turn with the screw and is fixed; the screws can pump water and generate power as turbine), and one true Archimedean screw (on the right, the cover is fixed to the screw and turns with it; this screw can only pump water)

There are a limited number of detailed studies on fish passage and Archimedes screws, most notably the study of (Schmalz 2010), who investigated wild local fish species including roach (*Rutilus rutilus*), bream (*Abramis brama*), eel (*Anguilla anguilla*), bullhead (*Cottus gobio*), three-spined stickleback (*Gasterosteus aculeatus*), spined loach (*Cobitis taenia*), and grayling (*Thymallus thymallus*), among others. In contrast to many claims that screws are inherently fishfriendly a substantial number of fish were found with scale loss, grinding injury, bleeding, and partial or complete cuts. In a study on the River Dart, UK, it was observed that almost all fish, including eels (*Anguilla anguilla*), trout (*Salmo trutta*) and salmonids (*Salmo salar*), passed through the Archimedes screw either unharmed (eels) or with negligible scale loss (salmon) (Kibel 2007, 2008; Brackely et al. 2018). Similarly, scale loss did not differ between treatment and control groups of salmon in a study on the River Don, Scotland (Brackely et al. 2016). However, the investigations of scale loss on euthanized individuals at the same site showed severe scale loss and distinctive patterns of scale loss due to grinding between the turbine blades and housing trough (Brackely et al. 2018). In addition, further studies found that fish with a body mass

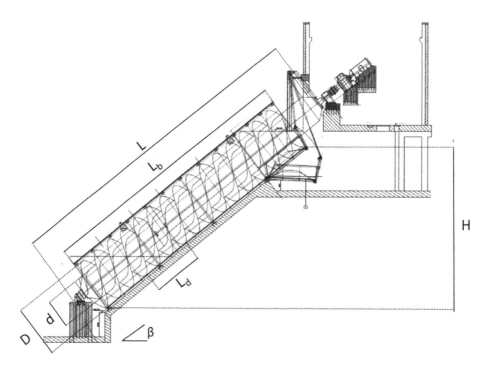

Fig. 11.2 Profile of the large Archimedes hydrodynamic screw at Ham (Belgium) that was studied within the FIThydro project, showing the injection location of fish and sensors on the top valve of the turbine (inset picture showing the valve in closed position; adapted from Pauwels et al. 2020)

less than 1 kg were not injured by contact with the screw leading edge if the tip speed was less than 4.5 m/s. The addition of a rubber leading edge further reduced injuries to larger fish at higher tip speeds (Kibel et al. 2009; Lyons and Lubitz 2013). In the study of river lamprey (*Lampetra fluviatilis*) on the River Derwent, UK, the damage rate was 1.5% for 66 juveniles released immediately upstream and who subsequently passed the Archimedes screw (Bracken and Lucas 2013). The impact of the screws in the River Sour, UK, and Diemel, Germany, were investigated by acoustically tagged eels (*Anguilla anguilla*) and salmon (*Salmo salar*). The behaviour of the eels in the River Sour was not found to be directly impacted by the screw passage. However, migration delay was introduced at this site by the fish being frequently milled and rejected back upstream (Piper et al. 2018). A screw study in the River Diemel observed a probability of 0–8% that a smolt would die after passing the screw (Havn et al. 2017). The findings of these studies show first that the published knowledge on Archimedes screws and fish passage are very limited in scope and are based on a limited number of live fish studies during Archimedes hydrodynamic screw passage. In order to improve designs, operational guidelines and improve downstream fish passage at screws, more research is needed to identify, define and establish the risk of injury and mortality to fish passing downstream through screws.

Apart from assessing the biological responses of live fishes, the development of safer screws can also be assessed by using passive sensors (Fig. 11.3). These sensors measure the physical conditions experienced during passage. Several studies for Kaplan turbines exist (Fu et al. 2016; Deng et al. 2005, 2010), however there is only a single study to date that has used sensors to measure the physical conditions in an Archimedes turbine (Boys et al. 2018). A recent sensor passage study evaluated event-based statistics including the number and severity of strike events, the nadir (lowest) and maximum pressures, and rate of pressure change. No live fish studies at the site were compared with sensor data in that study (Pauwels et al. 2020). However, two studies have combined live fish and sensor experiments (Deng et al. 2005, 2010). The first study was performed in a laboratory setting investigating shear-related injury and mortality, and the second related the percentage of severe events (collision and/or shear) to 48 h delayed mortality from live fish studies

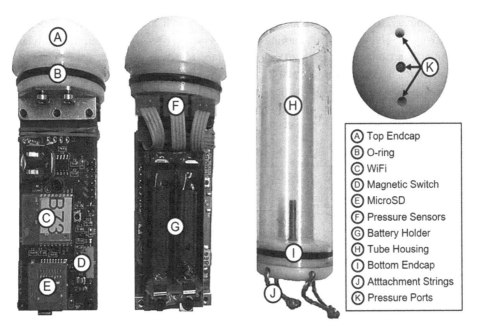

Fig. 11.3 Overview of the BDS sensors used in the study in the FIThydro project. The top endcap (**a, b**) contains three pressure transducers (**f, g**). Below there are two electronics boards containing the WiFi module (**c**), magnetic switch (**d**), microSD storage (**e**), and AAA battery holder (**g**). The sensor and electronics payload (**a-g**) is screwed by hand onto the bottom endcap (**i**), which also includes two rugged nylon attachment strings (**j**) for the balloon tags to bring the neutrally buoyant sensor back to the water surface (Pauwels et al. 2020)

in two Kaplan turbines. Therefore, the link between actively swimming fish and passive sensors remains to be conclusively investigated. Differences in the observed injury and mortality between fish species require multispecies, live fish experiments. Understanding the relationships among various strike variables and injury and mortality rates are necessary for improvements in turbine design), (Boys et al. 2018; Čada 2001). In our study within the FIThydro project, we evaluated injury and mortality of 2700 fish of three species that passed the Archimedes hydrodynamic screw of Ham shown in Fig. 11.2 at one of three rotational speeds: 30, 40 and 48 Hz. Additionally, we measured the total water pressure, linear acceleration, rotation rate, magnetic field intensity and absolute orientation (roll, pitch and yaw angles) during passage on each of the three rotational speeds with passive Barotrauma Detection System (BDS) sensors. The sensors illustrated in Fig. 11.3 were developed by the TalTech Centre for Biorobotics as part of the EU H2020 FIThydro project. We learned from this study that the chance to be injured or killed by the screws depends on the species. Substantial loss of fish due to screw passage was observed for bream, also for roach but not for eel, see Fig. 11.4. A screw passage was found to be a chaotic event and the relation between injuries and mortality and the rotational speed of the screw was not straightforward.

In summary, to date, the available studies strongly indicate that (A) Archimedes and hydrodynamic screws used as turbines are very unlikely to cause barotrauma-related fish injury and mortality, (B) mortality and injury rates are generally lower compared to conventional turbine types, but (C) they may cause injury and mortality, which is highly dependent on the fish species. Therefore, we stress the need for further studies on Archimedes screws to identify the causes of the observed species-specific injury

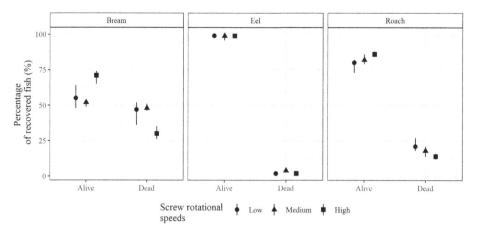

Fig. 11.4 Proportions of bream (Abramis brama), eel (Anguilla anguilla), and roach (Rutilus rutilus) indicating the state as either alive or dead after forced Archimedean screw turbine passage for each of the three rotational speeds tested: 33, 40 and 48 Hz (Pauwels et al. 2020)

and mortality rates. The largest challenge is to identify which screw characteristics significantly affect the rates of injury and mortality. Is it blade edge grinding, large-scale turbulence, shear stress, intermittent blade contact in the buckets or perhaps impingement between the blades and outer housing in hydrodynamic screws? Governments, policy makers, river managers and turbine designers need a list of the causes to design, build and remediate screws, to ensure that they provide a truly fishfriendly installation at each site. We believe these answers might specifically help to improve the design of larger screws (up to 10 MW) such as the one investigated in our study. Because screws can also pump water, improving their fish-friendliness, could make them better competitors for conventional pumps and turbines. To begin to address these key questions, it is imperative that future studies provide a list of standardized descriptions and physical metrics to cross-compare screws and identify the potential causes as they relate to the particular characteristics of the screw. We have provided an example of the basic characteristics needed for future studies in and have illustrated them on the profile of the investigated screw of Ham (Belgium; Fig. 11.2 and Table 11.1).

Acknowledgements This study was partially commissioned by the operator of the plant: De Vlaamse Waterweg NV. We specifically thank Koen Maeghe and Werner Dirckx from De Vlaamse Waterweg NV for the availability of the HPP facilities and for their continuous support.

Table 11.1 Basic characteristics needed in studies for the cross-comparison of Archimedes hydrodynamic screws, illustrated using values of the screw investigated in Ham, Belgium. Figure 11.2 indicates the screw parameters on the profile of the Ham screw (Pauwels et al. 2020)

Screw Parameters	Abbreviation	Value
Maximum power (MW)	–	1.2
Center tube length (m)	L	23
Helix length (m)	L_b	21.5
Slope (°)	β	38
Number of blades	–	3
Helix lead (m)	L_d	4.3
Centre tube diameter (m)	d	2.4
Helix diameter (m)	D	3.1
Helix operation (rpm/Hz/m³/s)	–	13.71/33/3 16.62/40/4 19.95/48/5
Gap between helix and housing (cm)	–	±2 cm
Fish deterrence system	–	None
Fish injury reduction measures	–	None

References

Boys C, Pflugrath BD, Mueller M, Pander J, Deng ZD, Geist J (2018) Physical and hydraulic forces experienced by fish passing through three different low-head hydropower turbines. Mar Freshw Res 69:1934–1944. https://doi.org/10.1071/MF18100

Bracken FSA, Lucas MC (2013) Potential impacts of small-scale hydroelectric power generation on downstream moving lampreys. River Res Appl 29:1073–1081. https://doi.org/10.1002/rra.2596

Brackley R, Bean C, Lucas M, Thomas R, Adams C (2016) Assessment of scale-loss to Atlantic salmon (*Salmo salar* L.) smolts from passage through an archimedean screw turbine. In Proceedings of the 11th ISE 2016, Melbourne, Australia

Brackley R, Lucas MC, Thomas R, Adams CE, Bean CW (2018) Comparison of damage to live v. euthanized Atlantic salmon *Salmo salar* smolts from passage through an Archimedean screw turbine. J Fish Biol 92:1635–1644. https://doi.org/10.1111/jfb.13596

Čada GF (2001) The development of advanced hydroelectric turbines to improve fish passage survival. Fisheries 26:14–23. https://doi.org/10.1577/1548-8446(2001)026%3c0014:tdoaht%3e2.0. co;2

Deng Z, Carlson TJ, Duncan JP, Richmond MC, Dauble DD (2010) Use of an autonomous sensor to evaluate the biological performance of the advanced turbine at Wanapum Dam. J Renew Sustain Energy 2:053104. https://doi.org/10.1063/1.3501336

Deng Z, Guensch GR, McKinstry CA, Mueller RP, Dauble DD, Richmond MC (2005) Evaluation of fish-injury mechanisms during exposure to turbulent shear flow. Can J Fish Aquat Sci 62:1513–1522. https://doi.org/10.1139/f05-091

Fu T, Deng ZD, Duncan JP, Zhou D, Carlson TJ, Johnson GE, Hou H (2016) Assessing hydraulic conditions through Francis turbines using an autonomous sensor device. Renew Energy 99:1244–1252. https://doi.org/10.1016/j.renene.2016.08.029

Havn TB, Sæther SA, Thorstad EB, Teichert MAK, Heermann L, Diserud OH, Borcherding J, Tambets M, Økland F (2017) Downstream migration of Atlantic salmon smolts past a low head hydropower station equipped with Archimedes screw and Francis turbines. Ecol Eng 105:262–275. https://doi.org/10.1016/j.ecoleng.2017.04.043

Kibel P (2007) Fish Monitoring and Live Fish Trials. Archimedes Screw Turbine, River Dart. Phase 1, Fishtek Consulting Ltd., Devon, UK

Kibel P (2008) Archimedes screw turbine fisheries assessment. phase II: eels and kelts. Aquac Eng 1:297–310

Kibel P, Pikc R, Coe T (2009) The Archimedes Screw Turbine: Assessment of Three Leading Edge Profiles. Fishtek Consulting Ltd., Devon, UK

Lashofer A, Hawle W, Pelikan B (2012) State of technology and design guidelines for the Archimedes screw turbine. https://tethys-engineering.pnnl.gov/publications/state-technology-design-guidelines-archimedes-screw-turbine (Access on 1 October 2012)

Lubitz WD, Lyons M, Simmons S (2014) Performance model of Archimedes screw hydro turbines with variable fill level. J Hydraul Eng 140:04014050. https://doi.org/10.1061/(ASCE)HY.1943-7900.0000922

Lyons M, Lubitz WD (2013) Archimedes screws for microhydro power generation. In Proceedings of the ASME 2013 7th International Conference on Energy Sustainability Collocated with the ASME 2013.In: Heat Transfer Summer Conference and the ASME 2013 11th International Conference on Fuel Cell Science, Minneapolis, USA

Pauwels IS, Baeyens R, Toming G, Schneider M, Buysse D, Coeck J, Tuhtan JA (2020) Multi-species assessment of injury, mortality, and physical conditions during downstream passage through a large archimedes hydrodynamic screw (Albert Canal, Belgium). Sustainability 12:8722. https://doi.org/10.3390/su12208722

Piper AT, Rosewarne PJ, Wright RM, Kemp PS (2018) The impact of an Archimedes screw hydropower turbine on fish migration in a lowland river. Ecol Eng 118:31–42. https://doi.org/10.1016/j.ecoleng.2018.04.009

Quaranta E, Revelli R (2018) Gravity water wheels as a micro hydropower energy source: A review based on historic data, design methods, efficiencies and modern optimizations. Renew Sustain Energy Rev 97:414–427. https://doi.org/10.1016/j.rser.2018.08.033

Schmalz W (2010) Untersuchungen zum Fischabstieg und Kontrolle Möglicher Fischschäden durch die Wasserkraftschnecke an der Wasserkraftanlage Walkmühle an der Werra in Meiningen—Abschlussbericht, Germany

Waters S, Aggidis GA (2015) Over 2000 years in review: revival of the archimedes screw from pump to turbine. Renew Sustain Energy Rev 51:497–505. https://doi.org/10.1016/j.rser.2015.06.028

Hydropeaking Impact Assessment for Iberian Cyprinids: Hydropeaking Tool Adaptation

12

Francisco Godinho⬤, Julie Charmasson⬤, Atle Harby⬤,
António Pinheiro⬤, and Isabel Boavida⬤

12.1 Introduction

Hydroelectric power plants operated in response to short-term, sub-daily changes of the electricity market, undergo rapid variations of turbine discharge, entailing quickly fluctuating water levels downstream (Moog 1993; Moreira et al. 2019). This operation regime, likely to rise in the near future in countries with increasing shares of variable renewable electricity generation (Ashraf et al. 2018), often causes numerous adverse impacts on river ecosystems, particularly fish assemblages (Moog 1993; Young et al. 2011; Schmutz et al. 2015).

F. Godinho (✉)
Hydroerg – Projectos Energéticos Lda, Lisbon, Portugal
e-mail: francisco.godinho@hidroerg.pt

J. Charmasson · A. Harby
SINTEF Energy Research, Trondheim, Norway
e-mail: julie.charmasson@sintef.no

A. Harby
e-mail: atle.harby@sintef.no

A. Pinheiro · I. Boavida
CERIS – Civil Engineering Research and Innovation for Sustainability, Instituto Superior Técnico /
University of Lisbon, Lisbon, Portugal
e-mail: antonio.pinheiro@tecnico.ulisboa.pt

I. Boavida
e-mail: isabelboavida@tecnico.ulisboa.pt

135
P. Rutschmann et al. (eds.), *Novel Developments for Sustainable Hydropower*,
https://doi.org/10.1007/978-3-030-99138-8_12

Although many rivers can naturally experience rapid flow changes, namely during floods, the hydrographs of peaking rivers are unique, leading to a harsh environment for freshwater organisms due to frequent and unpredictable disturbances, with no natural analogue (Poff et al. 1997; García et al. 2011). The hydrograph of peaking rivers can be characterized by parameters that change over space and time, such as magnitude, rate of change, frequency, duration, and timing (Harby and Noack 2013). Each of these parameters may be correlated with ecological consequences and therefore may be used to scale the impacts of hydropeaking.

The response of salmonid fishes to hydropeaking has been studied for some time (e.g. Valentin et al. 1996; Scruton et al. 2008; Puffer et al. 2015; Rocaspana et al. 2019). Salmonids can be affected by peaking, whereby the most common responses include stranding, drift, and dewatering of spawning grounds, which have been related to up- and down-ramping rates (Saltveit et al. 2001), peak flow magnitude (Auer et al. 2017), and baseflow duration (Casas-Mulet et al. 2016). In contrast, information is much scarcer regarding other fish taxa (e.g. Boavida et al. 2015, 2020), making it difficult to appraise peaking impacts of existing and new hydropower plants. Information gaps about hydropeaking impacts are particularly critical in the Iberian Peninsula, where threatened non-salmonid fish assemblages with high levels of endemicity coexist with existing and planned hydropower plants, including multi dam large hydropower schemes such as the one being constructed in the Tâmega river basin (Douro catchment). The Iberian freshwater fish fauna is characterized by native cyprinids that dominate river fish assemblages except for headwater streams and lowland rivers (Doadrio 2001; Reyjol et al. 2007). Given this scenario, it would be useful to have a tool in the Iberian Peninsula to quickly assess a priori hydropeaking impacts and to screen candidate hydropower plants for further investigations and for the implementation of appropriate mitigation measures.

Harby et al. (2016) developed a systematic approach in Norway to assess the additional impacts of hydropeaking on salmonid fish. The approach divides the impact from hydropeaking into two components: (direct) effects and vulnerability. The effect component characterises the possible impacts of peaking from how ecological relevant physical conditions changes, given the hydropower system and river morphology considering the regulated river as reference, whereas the vulnerability component characterise how vulnerable the system is to the additional impact from peaking.

Although the ecology of cyprinids is distinct from salmonid´s, this study adapts the tool for native Iberian cyprinids. The adaptation builds on the experience gathered so far on impacts of hydropeaking in Iberia (Boavida et al. 2015, 2020; Costa et al. 2019; Moreira et al. 2020; Oliveira et al. 2020) and on expert knowledge.

12.2 The Hydropeaking Tool for Salmonids

In the approach developed by Harby et al. (2016), physical conditions characterizing peaking flows consider the rate of flow change (water level change ratio), the dewatered area (change in water-covered area when flow is reduced from Qmax to Qmin), the magnitude of flow changes (Qmax/Qmin), and the frequency, timing and distribution of peaking operations.

For salmonid vulnerability, the following factors are taken into account in the approach: number of adult females, amount and distribution of spawning grounds, low flow periods, habitat degradation, low temperature impacts, pollution and other external factors. These effect and vulnerability factors are assessed for each HPP and are classified in semi-quantitative classes according to criteria developed from the literature, research in CEDREN (Centre for Environmental Design of Renewable Energy) or by expert opinion. The factors for peaking operations and vulnerability are finally combined to produce an overall assessment of hydropeaking impact at a particular site (from very high to small).

12.3 Adapting the Tool for Iberian Cyprinids

The general framework of the tool developed for salmonids was kept for the Iberian cyprinids, with the combined use of effect and vulnerability factors to assess the overall hydropeaking impact.

As initial step, a set of effect and vulnerability hydropeaking related factors/indicators were developed for Iberian cyprinids based upon available, published and unpublished, information. Moreover, preliminary thresholds separating different impact and vulnerability classes were established for each indicator to account for different levels of impact of hydropeaking on the focus taxa (the cyprinids *Luciobarbus bocagei*, *Pseudochondrostoma duriense*, *Squalius* spp. and *Achondrostoma* spp.).

The proposed factors/indicators and thresholds were then evaluated by eight experts on Iberian cyprinids ecology and Mediterranean rivers functioning. A final set of effect and vulnerability parameters/indicators was developed for Iberian cyprinids by including the expert opinions in the initial proposal (Tables 12.1 and 12.2).

All the effect parameters proposed for the salmonids were retained for the Iberian cyprinids, except the magnitude of flow changes, as assessed by Qmax/Qmin. Due to the limitations in available information, only three classes were established for each indicator when compared with the salmonids tool. Other differences included the consideration of distinct critical periods as well as different thresholds to classify some indicators given the specificity of the Iberian climate. Given the more generalist autoecology of the common Iberian cyprinids, the thresholds proposed were generally less stringent than the ones proposed for the salmonids.

Table 12.1 Final effect factors, indicators and criteria for characterization of Iberian non-salmonid rivers affected by hydropeaking

Effect factors	Indicator	Criteria for characterization		
		Very large (value 3)	Moderate (value 2)	Small (value 1)
E1: Rate of change	Water level change ratio (cm/h)	>15	15–5	< 5
E2: Dewatered area	Change in water-covered area when flow is reduced from Q_{max} to Q_{min} (%)	>40	40–10	< 10
E3: Frequency	Annual frequency (proportion/number of days per year with peaking)	> 75% (>273 d)	25–75% (91–273 d)	< 25% (<91 d)
E4: Distribution		Irregular during Spring (spawning period)	Irregular	Regular throughout the year
E5: Timing	Flow reductions in critical periods	During the spawning	During the Winter	During the low flow period

As expected, more differences are noticeable between the salmonids and the cyprinids vulnerability factors. In contrast to salmonids, two taxa groups were initially established, considering the Iberian barbel (*Luciobarbus bocagei*), the largest native species present in many Iberian rivers (e.g. Godinho et al. 1997), in one group, and the remaining cyprinids in another.

Instead of using the number of females as an indicator of the population size, the use of capture-per-unit-of-effort (CPUE; number of specimens collected in Spring with single-pass electrofishing /100 m^2) was proposed as indicator of abundance for the species or group of species considered. Initial threshold criteria to separate vulnerability classes were obtained as percentiles of the CPUE for barbel and the other cyprinids occurring in several Portuguese central and northern river reaches, including both natural and impacted reaches (The authors, unpublished data).

As a measure of recruitment limitations, the proportion of juvenile native cyprinid specimens, based on total length, are used, instead of the amount and distribution of spawning grounds considered for salmonids. Although growth for a particular species varies among different rivers and reaches, the following general size thresholds (total length, in mm) are proposed to identify juvenile specimens: Luciobarbus bocagei (120 mm); *Pseudochondrostoma duriense* and *Squalius carolitertii* (80 mm); *Squalius*

Table 12.2 Final vulnerability factors, indicators and criteria for characterization

Vulnerability factor	Indicator	Criteria for characterization		
		High (value 3)	Moderate (value 2)	Low (value 1)
V1a: Population size of native barbel (*Luciobarbus bocagei*)	Abundance: Capture-per-unit-of-effort (CPUE—number of specimens collected in Spring with single-pass electrofishing/100 m²)	≤1.5[1]	1.6–6.0[2]	>6.0
V1b: Population size of straight mouth nase (*Pseudochondrostoma* spp.)	Abundance: Capture-per-unit-of-effort (CPUE—number of specimens collected in Spring with single-pass electrofishing /100 m2)	≤2.0[3]	2.1–6.2[4]	>6.2
V1c: Effective population size of sensitive small native cyprinids (*Squalius alburnoides, Squalius carolitertii* and other *Squalius* spp.)	Abundance: Capture-per-unit-of-effort (number of specimens collected in Spring with single-pass electrofishing/100 m2)	≤1.5[5]	1.6–8.3[6]	>8.3
V2: Degree of limitations in recruitment	Proportion of juvenile native cyprinid specimens in Spring samples (based on specimens' length)	<30%	30–50%	50%-70%
V3: Habitat heterogeneity	River Habitat Survey (RHS, in Portugal) or the Spanish protocol for hydromorphological (HYMO) characterization of rivers (in Spain)	RHS or HYMO indicator compatible with bad ecological status	RHS or HYMO indicator compatible with moderate or mediocre status	RHS or HYMO indicator compatible with high or good status
V4: Habitat degradation	Change in magnitude and frequency of natural flood events	No floods	Some floods compared to the natural situation	Most of the natural floods still occur

[1] 30% percentile of the CPUE for nase occurring in 256 central and northern river reaches.

[2] 30% percentile of the CPUE for nase occurring in 256 central and northern river reaches.

[3] 30% percentile of the CPUE for nase occurring in 256 central and northern river reaches.

[4] 60% percentile of the CPUE for nase occurring in 256 central and northern river reaches.

[5] 30% percentile of the CPUE of small sized Iberian cyprinids (including *Squalius alburnoides* and *Squalius caroliterti*) occurring in 272 central and northern river reaches.

[6] 60% percentile of the CPUE of small sized Iberian cyprinids (including *Squalius alburnoides* and *Squalius carolitertii*) occurring in 272 central and northern river reaches.

alburnoides and *Achondrostoma* spp. (45 mm). The proposed values are a compromise between the maturity lengths for males and females. Habitat degradation was also included and assessed similarly as for salmonids, as the change in magnitude and frequency of natural flood events.

Low flow periods as bottleneck for salmonid fish stock size were not considered due to the tolerance of most Iberian cyprinids to low flow periods (e.g. Pires et al. 1999, 2010). The influence of reduced water temperature was also not included as a vulnerability factor. As for factors other than hydropeaking influencing the vulnerability of fish, a measure of habitat heterogeneity was included for Iberian cyprinids, since fish populations should be more vulnerable at homogeneous river reaches. Finally, the proportion of impacted river length compared to the total length was used for cyprinids as for the salmonids. This implies that we assume fish had access to the whole river length before hydropower development.

The joint assessment of the effect and vulnerability parameters was defined by adapting the combined assessment made for salmonids in Norway (Harby et al. 2016).

All the effect and vulnerability parameters were considered equally important and the values assigned to each one (from High, value 3, to Low, value 1) were added. The total scores for the effect and vulnerability parameters were then divided in three classes. For the parameter V1a, V1b and V1c a single value correspondent to the average of the species/species group naturally occurring in the river reach should be considered. In the end, an overall assessment of hydropeaking impact is made, by combining the effects of hydropeaking with the vulnerability of the river system (Fig. 12.1).

12.4 Discussion

Despite the different hydrographs between Nordic and Iberian rivers, most of the effect factors included in the initial tool were kept for Iberian rivers. This likely reflects the similar nature of hydropeaking irreflective of river type, in what it relates to inflow variations over space and time. From all the effect factors included for salmonids, the magnitude

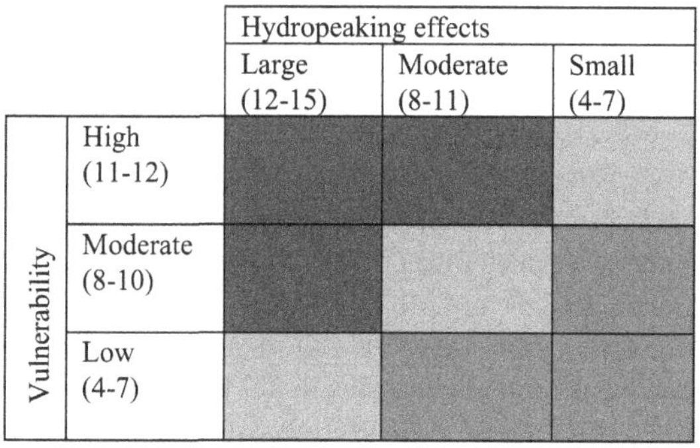

Fig. 12.1 Assessment matrix combining hydropeaking effects and vulnerability for total impact assessment. The colours denote the impact classes (large, moderate and small impacts are denoted, respectively, by red, yellow and green)

		Hydropeaking effects		
		Large (12-15)	Moderate (8-11)	Small (4-7)
Vulnerability	High (11-12)			
	Moderate (8-10)			
	Low (4-7)			

of flow changes was not kept for the Iberian rivers. The computation of this factor, as assessed by Qmax/Qmin, would invariably return larger values than for Norwegian HPPs since flow is near zero during the low flow period in many rivers in Mediterranean climate regions. The natural flow regime of Mediterranean-type streams is characterized by large differences between minimum and maximum discharge that are related with predictable, seasonal events of flooding and drying over an annual cycle (Gasith and Resh 1999; Bonada and Resh 2013).

Overall, the final set of effect factors was similar after the expert inputs, but some class thresholds were changed, namely for the dewatered area and the hydropeaking frequency. The distribution of hydropeaking events was also changed, with the highest impact linked to events occurring irregularly during Spring instead of irregular events occurring during all year. Spring was selected as a particularly vulnerable period as all Iberian cyprinids spawn largely during this season (e.g. Rodriguez-Ruiz and Granado-Lorencio 1992). In addition, regular hydropeaking events were considered less impacting, as fish individuals appear to memorize spatial and temporal environmental changes and to adopt a "least constraining" habitat (Halleraker et al. 2003; Costa et al. 2018; Jesus et al. 2019). Hydropeaking timing was also changed after the expert's input, with the highest impact related not only to the spawning period but also the sequent period of larvae development. In contrast to salmonids, density-related mortality during larvae period is unlikely for cyprinids, with year-class strength being related to stochastic environmental factors (Mills and Mann 1985). Consequently, hydropeaking could be particularly distressful for larvae in years where environmental conditions result in weak cohorts. The impact was considered reduced when occurring during the winter, and moderate if happening during the summer low flow period.

Contrasting with the effect factors, vulnerability factors for the cyprinids showed more differences with the ones proposed for the salmonids. These differences reflected the distinct auto-ecology of the two ray-finned fish families. First, we selected two taxonomical groups (Iberian barbel and smaller cyprinids), but based on expert's opinions, the breath of the smaller cyprinids justified the separation in two groups, one including the nase, and the other including the remaining cyprinids, but without *Achondrostoma* spp., due to their tolerance to hydropeaking and other anthropogenic impacts (Oliveira et al. 2012). The straight-mouth nases are usually the second largest cyprinid in fish assemblages, performing potamodromous spawning migrations such as the ones described for the barbel (Rodriguez-Ruiz and Granado-Lorencio 1992).

Since the number of females used for salmonids are more appropriate for an anadromous species such as the Atlantic salmon (*Salmo salar*) than for cyprinids, we opted to use CPUE as an indicator of population size of cyprinids. The abundance thresholds developed in this study were supported on available data on CPUE of native cyprinids in river reaches, but the indicator can be adapted to other databases on fish abundance, and can be also derived for specific river types. In the tool for salmonids, the rate of change (E1) is multiplied with the dewatered area (E2) factors. This is because the rate of change is not considered important if it does not lead to a significant reduction in dewatered area when water levels sink, and vice versa. This is due to the risk of stranding, which is considered a major challenge for salmonids. In our system, the effect factors are just an addition of all factors, because other impacts like disturbing movements, changing habitats, access to feeding, spawning, are also equally important. Besides, dewatered areas in Mediterranean-streams are typically large due to peak magnitude.

The hydropeaking tool developed for salmonids in Norway was successfully adapted to Iberian cyprinids and Mediterranean rivers. Nevertheless, it should be emphasized that both effect and vulnerability factors and the criteria for their characterization might be improved in the future if new studies on Iberian hydropeaking rivers come out with new insights.

References

Ashraf F, Haghighi AT, Riml J, Alfredsen K, Koskela JJ, Kløve B, Marttila H (2018) Changes in short term river flow regulation and hydropeaking in Nordic rivers. Sci Rep 8:17232. https://doi.org/10.1038/s41598-018-35406-3

Auer S, Zeiringer B, Führer S, Tonolla D, Schmutz S (2017) Effects of river bank heterogeneity and time of day on drift and stranding of juvenile European grayling (Thymallus thymallus L.) caused by hydropeaking. Sci Total Environ 575:1515–1521. https://doi.org/10.1016/j.scitotenv.2016.10.029

Boavida I, Ambrósio F, Costa MJ, Quaresma A, Portela MM, Pinheiro A, Godinho F (2020) Habitat use by *Pseudochondrostoma duriense* and *Squalius carolitertii* downstream of a small-scale hydropower plant. Water 12(9):2522

Boavida I, Santos JM, Ferreira T, Pinheiro A (2015) Barbel habitat alterations due to hydropeaking. J Hydro-Environ Res 9:1570–6443. https://doi.org/10.1016/j.jher.2014.07.009

Bonada N, Resh VH (2013) Mediterranean-climate streams and rivers: geographically separated but ecologically comparable freshwater systems. Hydrobiol 719:1–29. https://doi.org/10.1007/s10750-013-1634-2

Casas-Mulet R, Alfredsen K, Brabrand A, Saltveit SJ (2016) Hydropower operations in groundwater-influenced rivers: Implications for Atlantic salmon. Salmo salar, early life stage development and survival. Fish Manag Ecol 23: 144–151. https://doi.org/10.1111/fme.12165

Costa MJ, Fuentes-Pérez JF, Boavida I, Tuhtan JA, Pinheiro AN (2019) Fish under pressure: Examining behavioural responses of Iberian barbel under simulated hydropeaking with instream structures. PLoS ONE 14(1): e0211115. https://doi.org/10.1371/journal.pone.0211115

Costa MJ, Boavida I, Almeida V, Cooke SJ, Pinheiro AN (2018) Do artificial velocity refuges mitigate the physiological and behavioural consequences of hydropeaking on a freshwater Iberian cyprinid? Ecohydrology, 11:e1983. https://doi.org/10.1002/eco.1983

Doadrio I (ed) (2001) Atlas y libro rojo de los peces continentales de España. Dirección General de Conservación de la Naturaleza, Ministerio De Medio Ambiente, Madrid, pp.364

Drescher M, Perera AH, Johnson CJ, Buse LJ, Drew CA, Burgman MA (2013) Toward rigorous use of expert knowledge in ecological research. Ecosphere 4(7):83. https://doi.org/10.1890/ES12-00415.1

García A, Jorde K, Habit E, Caamaño D, Parra O (2011) Downstream environmental effects of dam operations: Changes in habitat quality for native fish species. River Res Appl 27:312–327. https://doi.org/10.1002/rra.1358

Gasith A, Resh VH (1999) Streams in Mediterranean climate regions: abiotic influences and biotic responses to predictable seasonal events. Annu Rev Ecol Syst 30:51–81. https://doi.org/10.1146/annurev.ecolsys.30.1.51

Godinho FN, Ferreira MT, Cortes RV (1997) Composition and spatial organization of fish assemblages in the lower Guadiana basin, southern Iberia. Ecol Freshw Fish 6:134–143. https://doi.org/10.1111/j.1600-0633.1997.tb00155.x

Harby A, Forseth T, Ugedal O, Bakken TH, Sauterleute J (2016) A Method to assess impacts from hydropeaking. 11th International Symposium on Ecohydraulics (ISE 2016), Melbourne, Australia. Extended Abstract

Harby A, Noack M (2013) Rapid flow fluctuations and impacts on fish and the aquatic ecosystem. In Maddock I, Harby A, Kemp P, Wood P (eds) Ecohydraulics: An Integrated Approach, First Edition. John Wiley & Sons, Ltd. https://doi.org/10.1002/9781118526576.ch19

Halleraker JH, Saltveit SJ, Harby A, Arnekleiv JV, Fjellstad HP, Kohler B (2003) Factors influencing stranding of wild juvenile brown trout (*Salmo trutta*) during rapid and frequent flow decreases in an artificial stream. J River Res Appl 19:589–603. https://doi.org/10.1002/rra.752

Jesus J, Teixeira A, Natário S, Cortes R (2019) Repulsive effect of stroboscopic light barriers on native salmonid (*Salmo trutta*) and cyprinid (*Pseudochondrostoma duriense* and *Luciobarbus bocagei*) species of Iberia. Sustain 11(5):1332

Mills CA, Mann RHK (1985) Environmentally-induced fluctuations in year-class strength and their implications for management. J Fish Biol 27:209–226. https://doi.org/10.1111/j.1095-8649.1985.tb03243.x

Moog O (1993) Quantification of daily peak hydropower effects on aquatic fauna and management to minimize environmental impacts. Regul Rivers Res Mgmt 8:5–14. https://doi.org/10.1002/rrr.3450080105

Moreira M, Hayes DS, Boavida I, Schletterer M, Schmutz S, Pinheiro A (2019) Ecologically-based criteria for hydropeaking mitigation: A review. Sci Total Environ 657:1508–1522. https://doi.org/10.1016/j.scitotenv.2018.12.107

Moreira M, Costa MJ, Valbuena-Castro J, Pinheiro AN, Boavida I. (2020) Cover or velocity: what triggers Iberian barbel (*Luciobarbus bocagei*) refuge selection under experimental hydropeaking conditions? Water 12(2):317. https://doi.org/10.3390/w12020317

Oliveira JM, Segurado P, Santos JM, Teixeira A, Ferreira MT, Cortes RV (2012) modelling stream-fish functional traits in reference conditions: Regional and local environmental correlates. PLoS ONE 7(9):e45787. https://doi.org/10.1371/journal.pone.0045787

Oliveira IC, Alexandre CM, Quintella BR, Almeida PR. (2020) Impact of flow regulation for hydroelectric production in the movement patterns, growth and condition of a potamodromous fish species. Ecohydrology, 13:e2250. https://doi.org/10.1002/eco.2250

Pires AM, Cowx IG, Coelho MM (1999) Seasonal changes in fish community structure of intermittent streams in the middle reaches of the Guadiana basin, Portugal. J Fish Biol 54:235–249. https://doi.org/10.1111/j.1095-8649.1999.tb00827.x

Pires DF, Pires AM, Collares-Pereira MJ, Magalhães MF (2010) Variation in fish assemblages across dry-season pools in a Mediterranean stream: Effects of pool morphology, physicochemical factors and spatial context. Ecol Freshw Fish 19:74–86. https://doi.org/10.1111/j.1600-0633.2009.00391.x

Poff NL, Allan D, Bain M, Karr JR, Prestegaard KL, Richter BD, Sparks RE, Stromberg JC (1997) The natural flow regime. A paradigm for river conservation and restoration. Bioscience 47:769–784. https://doi.org/10.2307/1313099

Puffer M, Berg OK, Huusko A, Vehanen T, Forseth T, Einum S (2015) Seasonal effects of hydropeaking on growth, energetics and movement of juvenile Atlantic Salmon (*Salmo Salar*). River Res Applic 31:1101–1108. https://doi.org/10.1002/rra.2801

Radinger J, Kail J, Wolter C (2017) Differences among expert judgments of fish habitat suitability and implications for river management. River Res Applic 33:538–547. https://doi.org/10.1002/rra.3109

Reyjol Y, Hugueny B, Pont D, Bianco PG, Beier U, Caiola N, Casals F, Cowx I, Economou A, Ferreira T, Haidvogl G, Noble R, DeSostoa A, Vigneron T, Virbickas T (2007) Patterns in species richness and endemism of European freshwater fish. Glob Ecol Biogeogr 16:65–75. https://doi.org/10.1111/j.1466-8238.2006.00264.x

Rocaspana R, Aparicio E, Palau-Ibars A, Guillem R, Alcaraz C (2019) Hydropeaking effects on movement patterns of brown trout (*Salmo trutta* L.). River Res Applic 35:646–655. https://doi.org/10.1002/rra.3432

Rodriguez-Ruiz A, Granado-Lorencio C (1992) Spawning period and migration of three species of cyprinids in a stream with Mediterranean regimen (SW Spain). J Fish Biol 41:545–556. https://doi.org/10.1111/j.1095-8649.1992.tb02682.x

Saltveit S, Halleraker J, Arnekleiv J, Harby A (2001) Field experiments on stranding in juvenile atlantic salmon (*Salmo salar*) and brown trout (*Salmo trutta*) during rapid flow decreases caused by hydropeaking. Regul Rivers Res Mgmt 17:609–622. https://doi.org/10.1002/rrr.652

Schmutz S, Bakken TH, Friedrich T, Greimel F, Harby A. Jungwirth M, Melcher A, Unfer G, Zeiringer B (2015) Response of fish communities to hydrological and morphological alterations in hydropeaking rivers of Austria. River Res Applic 31: 919– 930. https://doi.org/10.1002/rra.2795

Scruton DA, Pennell C, Ollerhead LMN, Alfredsen K, Stickler M, Harby A, Robertson M, Clarke KD, LeDrew LJ (2008) A synopsis of 'hydropeaking' studies on the response of juvenile Atlantic salmon to experimental flow alteration. Hydrobiol 609:263–275. https://doi.org/10.1007/s10750-008-9409-x

Valentin S, Lauters F, Sabaton C, Breil P, Souchon Y (1996) Modelling temporal variations of physical habitat for brown trout (*Salmo trutta*) in hydropeaking conditions. Regul Rivers Res Mgmt 12:317–330. https://doi.org/10.1002/(SICI)1099-1646(199603)12:2/3%3c317::AID-RRR398%3e3.0.CO;2-1

Young PS, Cech JJ, Thompson LC (2011) Hydropower-related pulsed-flow impacts on stream fishes: a brief review, conceptual model, knowledge gaps, and research needs. Rev Fish Biol Fisheries 21:713–731. https://doi.org/10.1007/s11160-011-9211-0

Ianina Kopecki, Matthias Schneider⬤, and Martin Schletterer⬤

13.1 Introduction

Hydropeaking alters hydraulic conditions as well as wetted areas and is disadvantageous for fish species with specific hydraulic preferences, in particular for spawning grounds and fish stages with restricted mobility such as juvenile fish and larvae (Moreira et al. 2019). Habitat models describe the environmental conditions for fish and use the requirements of fish related to these conditions to calculate habitat suitability. Moreover, habitat modelling is also an appropriate tool to quantify the impacts of hydropeaking since these impacts can be interpreted as a decrease of habitat suitability. In contrast to the standard habitat parameters as water depth, flow velocity and granulometry, the temporal change of habitat conditions and the speed of this temporal change are highly relevant for hydropeaking analyses. The habitat model system CASiMiR has been extended by a hydropeaking module to take account of these impacts (Schneider and Kopecki 2016). More precisely the following features have been integrated:

I. Kopecki · M. Schneider (✉)
sje – Ecohydraulic Engineering GmbH, Stuttgart (Vaihingen), Germany
e-mail: mailbox@sjeweb.de

I. Kopecki
e-mail: kopecki@sjeweb.de

M. Schletterer
Department of Hydropower Engineering, Group Ecology, TIWAG – Tiroler Wasserkraft AG, Innsbruck, Austria
e-mail: martin.schletterer@tiwag.at; martin.schletterer@boku.ac.at

Institute of Hydrobiology and Aquatic Ecosystem Management, University of Natural Resources and Life Sciences, Vienna, Austria

© The Author(s) 2022
P. Rutschmann et al. (eds.), *Novel Developments for Sustainable Hydropower*,
https://doi.org/10.1007/978-3-030-99138-8_13

- downramping rates, stranding risk and spatial integration
- upramping rates, drift risk and spatial integration
- redd stability
- habitat shift and habitat persistence

These model features serve on one hand as tools for the detailed hydraulic-based assessment of hydropeaking scenarios and mitigation measures. On the other hand, they provide input parameters for a matrix-based assessment of hydropeaking impact strength (Boavida et al. 2020).

13.2   Testcase GKI (Inn, Austria)

The upper Inn River in Tyrol is affected by hydropeaking from the hydropower scheme Pardella-Martina in Switzerland (Meier 1991). The diversion hydropower plant GKI (Gemeinschaftskraftwerk Inn) has been designed to mitigate hydropeaking along the stretch between the villages Ovella and Prutz (Herdina 2018). This Testcase enabled the comparison between the impacted and mitigated situation. The newly developed CASiMiR hydropeaking module, covering different risks for fish habitats arising from the rapid flow changes has been applied in the Testcase GKI to quantify and assess the ecological impacts of different hydropeaking events. Analyses were carried out in three morphologically different river stretches: the homogeneous channel-like reach Kajetansbrücke, the more heterogeneous reach Mariastein 1 with alternating gravel bars and the most heterogeneous reach Mariastein 2 with characteristics of a braided river (see Fig. 13.1).

For all three reaches unsteady 2D hydrodynamic models have been set up with HYDRO_AS-2D. These are the basis for the CASiMiR analysis and deliver detailed information for some of the before mentioned effect factors such as water level change rates and wetted areas, but also further parameters such as maximum flow velocities and sediment movement.

13.3   Identifying Representative Hydropeaking Events

For the peak impact assessment, sensitive periods for specific life stages of grayling were considered. March to April covers the grayling spawning and brown trout larvae season, whereas May to June is the emergence and early larvae period of grayling. October and November again describe the spawning season of brown trout. The flow time series in Fig. 13.2 reflects the temporal discharge variation at the gauging station Kajetansbrücke in the investigation stretch. The increase in the discharge for the months of May and July

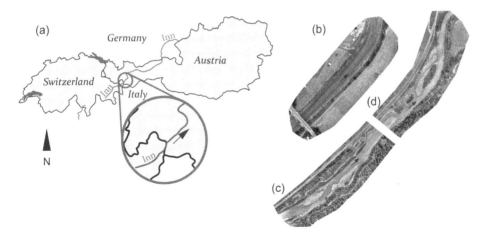

Fig. 13.1 **a** Location of the hydropower plant GKI; Aerial pictures of the selected investigation stretches: **b** Kajetansbrücke, **c** Mariastein 1 and **d** Mariastein 2

is caused mainly from snowmelt in the Alps. Summer months do not only present higher flows, but also higher variability of discharge.

Main outcomes of the analysis are that the May/June period shows characteristics that are different to March/April and Oct/Nov. The base flow is not as constant as in the other periods, high fluctuations between 50 and 150 m^3/s are detected, small events with a low baseflow are not present (see Fig. 13.2a). In consequence, the mentioned three periods were analysed statistically to derive representative events for the CASiMiR analysis. Not necessarily an average event or the most extreme event is suitable to describe the impact but rather events with high amplitudes and high change rates that occur regularly. Similar as in the COSH tool (Sauterleute and Charmasson 2014), after statistical evaluation, percentiles can be used to define this kind of events. The identification of peak events was made using a Peak Detection Model applying several steps. These steps and their relevance are listed in Table 13.1.

Using the Peak detection model, events with different start- and end-flows and different up- and downramping rates could be identified as typical for certain periods of the year. Based on this analysis, for the current (impacted) situation, five different peaking events were detected and used for further calculations with CASiMiR:

Event 1: Big event [5 ➔ 94 m^3/s] (Mar/April and Oct/Nov)
Event 2: Small event [5 ➔ 49 m^3/s] (Mar/April and Oct/Nov)
Event 3: Big event with low base flow [20 ➔ 97 m^3/s] (May/June)
Event 4: Small event with low base flow [16 ➔ 46 m^3/s] (May/June)
Event 5: Big event with high base flow [68 ➔ 130 m^3/s] (May/June)

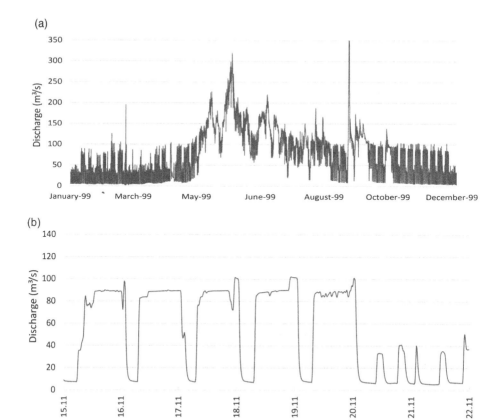

Fig. 13.2 Hydrological regime at Kajetansbrücke **a** 1999 hydrograph, **b** representative week hydrograph for a winter half-year

Table 13.1 Steps of hydropeaking event detection

Smoothing	Prevents, that small variations of flow rate within an event are detected as a peak event
Ramping rate	Prevents, that flow changes with small ramping rates are detected as a peak event
Difference to neighbouring flows and minimum ramping rate	Prevent the subdivision of one event into several due to small changes of the ramping rate within an event
Minimum Duration	Prevents, that short term variations are detected as an event
Merging of events	Prevents, that continuous events with small interruptions are detected as separate events

For these events, we analysed:

(a) Stranding risk

The stranding risk for juveniles of grayling and brown trout was calculated using a fuzzy rule-based approach. Combinations of "low" water depth and "high" water level change rate lead to "high" stranding risk. "High" water depth does not imply any risk, independent of water level change rate. Thresholds for "high", "medium" and "low" risk were defined with values lower 12 cm/h being uncritical, between 12 and 30 cm/h being critical and higher than 30 cm/h being very critical (Schmutz et al. 2013). Results for reach Kajetansbrücke and reach Mariastein 1 and event #3 are shown in Fig. 13.3 for grayling, in terms of a Risk Index RI between 0 (low) and 1 (high).

To gain quantitative information for the risk in the whole reaches the model elements with increased risk were integrated in terms of a Weighted Risk Area

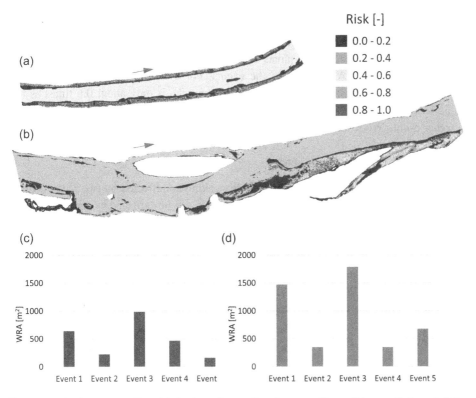

Fig. 13.3 Maximum stranding risk for juvenile grayling for event #3 conditions: **a** Kajetansbrücke, **b** Mariastein 1″. Weighted stranding risk area for juvenile grayling: **c** Kajetansbrücke, **d** Mariastein 1

(WRA) equivalent to the Weighted Usable Area WUA used in Habitat suitability investigations (Bovee and Cochnauer 1977).

$$WRA = \sum_{i=1}^{n} A_i \cdot RI_i$$

A_i = area of model element I, RI_i = Risk Index of model element i.

The WRA for both reaches are shown for all 5 representative peaking events in Fig. 13.3. The WRA for juvenile grayling during event #3 is about 1.7 times higher in reach Mariastein 1 than in reach Kajetansbrücke. However, this is nearly the relation of the wetted area in both reaches, so the proportion of risk area is about the same. In reach Kajetansbrücke the risk index RI is higher in some locations, but there are also extended areas with low risk available that are usable for larvae.

(b) Drift risk

In CASiMiR drift risk is calculated by a fuzzy rule-based approach that considers combinations of "low" water depth and "high" flow velocity as critical. Results in Fig. 13.4 indicate that high-risk areas for juvenile grayling are larger in reach Kajetansbrücke and that low-risk areas appear in reach Mariastein 1 in much higher quantities. The WRA depends on the events. For event #3 the WRA is about the same in both reaches, but since reach Kajetansbrücke is only about 0.6 times as big as reach Mariastein 1 the risk potential in the first one is higher. However, for the other events WRA in the second reach is partly higher than in the first reach.

(c) Spawning habitats and persistence

Spawning areas in river reaches with hydropeaking are affected by two factors. First, they can fall dry during low flow and second, eggs can be damaged by sediment movement during high flows. Figure 13.5 shows the suitability of theoretically available persistent spawning grounds during the whole peaking event #3 (minimum suitability for all time steps) together with the areas where suitable spawning substratum (size approx. 16–32 mm) gets in to motion during increasing flow (dashed areas). Our analyses revealed that (a) almost no areas are available in the analysed stretches that have a suitability higher than 0.2 (persistent spawning areas) and (b) large areas would be affected by movement of suitable spawning substratum, if present (currently not the case).

(d) Habitat shift

Finally, the spatial shift of suitable habitats for juvenile fish is a hazard for grayling and other fish. Due to their reduced mobility, juveniles up to a certain age cannot overcome larger distances when their habitats are moving with the water edge. This

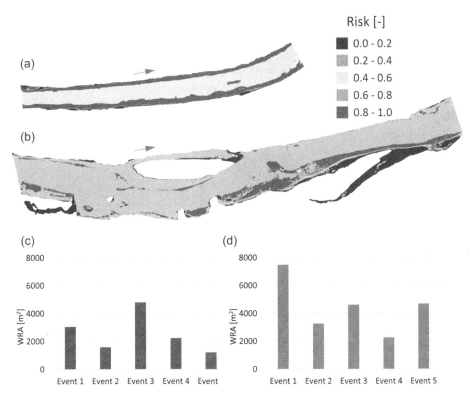

Fig. 13.4 Maximum drift risk for juvenile grayling for event #3 conditions: **a** Kajetansbrücke, **b** Mariastein 1. Weighted drift risk area for juvenile grayling: **c** Kajetansbrücke, **d** Mariastein 1

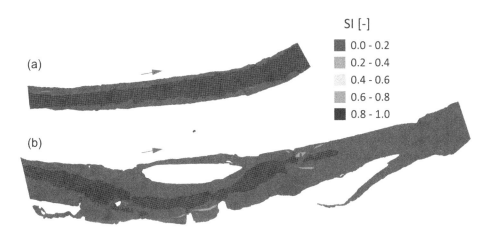

Fig. 13.5 Persistent spawning areas for grayling during event #3: **a** Kajetansbrücke, **b** Mariastein 1

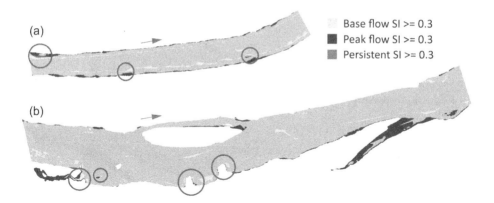

Fig. 13.6 Displacement (shift) of suitable habitats for juvenile grayling between based and peak flow for Event #3 conditions: **a** Kajetansbrücke, **b** Mariastein 1

risk can be assessed by visualizing the location of good juvenile habitats for base flow and peak flow and overlaying them.

Figure 13.6 shows the location of good juvenile habitat for both flow situations during event #3. In some locations (marked with green circles) the distance between good habitats for base flow and peak flow is comparatively small. These are the habitat shifts, which larvae can presumably follow when experiencing a peak event.

13.4 Conclusions and Outlook

Currently the Upper Inn River is affected by hydropeaking. Along the stretch between Ovella and Prutz we have exemplified the application of the CASiMiR hydropeaking module. This module allows to analyse habitat availability as well as the risk of stranding and drift (of fish larvae) during different hydropeaking events. It enables the comparison of different planning scenarios and can serve as an assessment and planning tool.

The GKI is Europe's first large hydropeaking diversion plant. Combining a buffer reservoir and a diversion stretch, it is possible to divert the hydropeaking further downstream to a larger catchment (Moreira et al. 2020), as foreseen by the water management framework plan Western Tyrol (Reindl et al. 2017).

The concept of the GKI will reduce hydropeaking in the Upper Inn between Ovella and Prutz, as the rapid habitat shifts are omitted (Herdina 2018; Moreira et al. 2020). The application of the CASiMiR hydropeaking module to the current situation and the situation after the implementation of GKI has confirmed that with the operation of the new HPP the hydropeaking impacts on fish will be significantly reduced and that the diversion of large hydropeaking events is an effective mitigation measure.

Acknowledgements Thanks to Johann Herdina (managing director of the Gemeinschaftskraftwerk Inn GmbH and member of the Management Board of TIWAG – Tiroler Wasserkraft AG) for providing this Testcase to the FIThydro project. We acknowledge Peter Mayr and Hansjörg Reiner (both: flussbau iC) for echosounding and settig up the hydraulic model. Further, we thank Max Boschi and Rudi Schneeberger for providing drone orthomosaics. Finally, we thank Luis Daniel Soto Molero for performing the analyses with the CASiMiR hydropeaking module, as well as António Pinheiro and Miguel Moreira (both: Instituto Superior Técnico, Universidade de Lisboa) for providing hydrological time series (hydropeaking scenarios) for the analyses.

References

Boavida I, Díaz-Redondo M, Fuentes-Pérez JF, Hayes DS, Jesus J, Moreira M, Belmar O, Vila-Martínez N, Palau-Nadal A, Costa MJ (2020) Ecohydraulics of river flow alterations and impacts on freshwater fish. Limnetica 39(1):213–232. https://doi.org/10.23818/limn.39.14

Bovee KD, Cochnauer T (1977) Development and evaluation of weighted criteria, probability-of-use curves for instream flow assessments: fisheries. In: Instream flow information paper 3. United States Fish and Wildlife Service FWS/OBS-77/63, p 38

Herdina J (2018) GKI hydroelectric power plant project: technical and contractual challenges. Tunnel 48–56. May 2018. Retrieved from: https://www.tunnel-online.info/en/artikel/tunnel_GKI_Hydroelectric_Power_Plant_Project_Technical_and_contractual_3246870.html

Meier R (1991) The new stage of Pradella-Martina of the Engadiner Kraftwerke AG. Die neue Stufe Pradella-Martina der Engadiner Kraftwerke. Bulletin des Schweizerischen Elektrotechnischen Vereins (und) des Verbandes Schweizerischer Elektrizitaetswerke 82(6):29–32

Moreira M, Hayes DS, Boavida I, Schletterer M, Schmutz S, Pinheiro A (2019) Ecologically-based criteria for hydropeaking mitigation: a review. Sci Total Environ 657:1508–1522. https://doi.org/10.1016/j.scitotenv.2018.12.107

Moreira M, Schletterer M, Quaresma A, Boavida I, Pinheiro A (2020) New insights into hydropeaking mitigation assessment from a diversion hydropower plant: the GKI project (Tyrol, Austria). Ecol Eng 158:106035

Reindl R, Egger K, Fitzka G (2017) Der Wasserwirtschaftliche Rahmenplan Tiroler Oberland. Wasser-Wirtschaft 7–8:69–74. (in German). Reindl R, Egger K, Fitzka G (2017) The water management framework for the Tyrolean Oberland. Wasser-Wirtschaft 7–8:69–74. (in English)

Schmutz S, Fohler N, Friedrich T, Fuhrmann M, Graf W, Greimel F, Höller N, Jungwirth M, Leitner P, Moog O, Melcher A, Müllner K, Ochsenhofer G, Salcher G, Steidl C, Unfer G, Zeiringer B (2013) Schwallproblematik an Österreichs Fließgewässern—Ökologische Folgen und Sanierungsmöglichkeiten. BMFLUW, Wien, p 176

Sauterleute JF, Charmasson J (2014) A computational tool for the characterisation of rapid fluctuations in flow and stage in rivers caused by hydropeaking. Environ Model Softw 55:266–278. https://doi.org/10.1016/j.envsoft.2014.02.004

Schneider M, Kopecki I (2016) Abbildung der Auswirkungen von Schwall und Sunk mit dem Habitatmodell CASiMiR. In: Schwall & Sunk: Forschungsstand & Ausblick (Umwelt, Schriftreihe für Ökologie und Ethologie—Sonderband). Facultas Universitätsverlag, pp 115–123 (in German)

Creation and Use of "Compensation" Habitats—An Integrated Approach

14

Georg Loy and Walter Reckendorfer

14.1 Introduction

To foster and conserve fish populations, large efforts in re-establishing connectivity and restructuring rivers have been made by the Hydropower sector. Most of these efforts aimed to pursue the common goal of reaching the good environmental potential according to the water framework directive (WFD) in heavily modified water bodies. According to the sectors experiences, a "best environmental option" should be implemented and chosen as a solution to reach the good ecological potential. The focus of the plant owners is to identify key habitats for specific species and establish these along the rivers, tributaries, oxbows and especially in river like bypass channels. The attempt is to offer accessible habitats for the whole live cycle of fish starting from spawning, to the juvenile phase up to the adult stage. The consideration of habitat issues is of uppermost importance for the conservation of fish populations. The restoration of longitudinal connectivity without consideration of habitat issues leads to sub-optimal solutions in river restoration.

A systemic approach should also be pursued in guidelines for fish passage. The focus on hydraulic design parameters in most of the available guidelines often results in technical solutions as it is easier to adhere to the hydraulic guide values implementing technical fishways such as vertical slot passes. The focus on fishway hydraulics also may shift attention from other important issues that must be addressed to reach the goals of the WFD,

G. Loy
VERBUND Innkraftwerke GmbH, Töging am Inn, Germany
e-mail: georg.loy@verbund.com

W. Reckendorfer (✉)
VERBUND Hydro Power GmbH, Vienna, Austria
e-mail: walter.reckendorfer@verbund.com

© The Author(s) 2022
P. Rutschmann et al. (eds.), *Novel Developments for Sustainable Hydropower*,
https://doi.org/10.1007/978-3-030-99138-8_14

such as habitat availability. The goal should be to foster all endangered rheophilic and lithophilic species and thus to identify and implement the fluviatile aspects necessary for their protection and promotion.

14.2 Ecological Measures in Impoundments of River Power Stations

At the Inn River in Germany and along the border to Austria, VERBUND is operating a cascade of run off the river hydropower plants. The first plant started operation in the beginning of the 19th century. However, the deterioration of the river started much earlier related to navigation, flood protection and the development of agricultural areas. The Inn River has been straightened and banks have been fixed, which enhanced incision of the river bed. The construction of hydropower plants also aimed to stabilize the river bed. Modifications like impounding, bank protection and further flood control works have completely changed the river system. The siltation processes and the high sand transport formed secondary flood plains and oxbows with a high primary productivity. Over time, major floods and the ongoing sedimentation left only minor structures with almost no lateral connectivity to oxbows.

To counteract these processes the operator VERBUND, together with authorities and planners developed a large scale restoration scheme for the Inn River in Germany. The focus of the first measures was along the impoundments of the hydropower plants Wasserburg, Teufelsbruck and Gars (start of operation in 1938) with the goal to identify measures to foster and protect existing fish species. The focus was not only on rheophilic potamodromous fish species, but also on the improvement of the fish biomass as fish play a major role in the existing bird sanctuaries especially with regard to birds prey. A decline of the fish population has been observed by local fisherman since the last three decades.

In a first step former and existing habitat elements were identified and classified to certain habitat needs of populations. A so-called "fish habitat concept" was the overall approach to identify possible measures. The aim was to identify the main habitat needs for all different stages of development. Starting from spawning, juvenile up to seasonal habitat use e.g. floods and winter conditions. A major challenge in the analysis was, that each hydropower plant and the respective river stretches provided differing natural constraints due to the local conditions (geometry) such as gorge type, or wide artificial wet lands (reed and willows) but also due to river training and the high fine sediment loads of the Inn River.

The German Federal Water Act (Wasserhaushaltsgesetz, WHG) demands concepts and measures to protect fish population as a requirement to operate hydropower plants. Large hydropower plants with low head and large Kaplan turbines have lower mortality rates than small plants. Additionally, technical solutions, in this scale, are not available (Reckendorfer et al. 2017). Further, existing solutions such as small trash rakes implemented at the existing inflow structures would either increase velocity or, in the case of the Inn River, with high sand transport and high loads of large driftwood, would make existing hydropower production almost impossible.

Thus, alternative approaches and measures for fish protection had to be developed and applied along the River Inn. After analysing the historical conditions and discussions with stakeholders and decision makers, VERBUND and a team of involved experts decided to implement as many key habitats to the system as possible and furthermore add certain key habitats into the bypass systems, i.e. the connectivity measures have been combined with several main habitat components (flow, gravel, shallow parts, ponds etc.). In large rivers such as the Inn most of the eco-morphological structures and components are difficult or almost impossible to reconstruct or maintain within the main river. It is known that key habitats show a substantial effect on fish population whereas technical measures such as smaller trash rakes lack their verification on a population level.

Flood protection necessities, land and forest use, ownership and sanctuaries are constraints to be identified prior to the concept phase of implementation of restoration measures. In the concept phase starting with the historic river system, the preconditions due to flood protection and impounding were identified, including also secondary floodplains with their ecological functions. These floodplains had a variety of warmer shallow lagoons connected to the main river. The possibility to add these former structures into the existing system were investigated by the project team. These secondary wetlands play a major role for the bird sanctuaries as feeding grounds due to their high primary and secondary productivity.

As gravel is not any longer transported through the series of plants a concept to compensate for the lack in gravel transportation and its function was also necessary. The main ecological functions of gravel for fish such as providing spawning grounds and nurseries can be provided in connected and restructured tributaries, at shores with removed bank protection and in newly created bypass rivers.

At different sections of a reservoir of a run-of-river plant different measures are identified (Holzner et al. 2014; Loy et al. 2014) and might be possible to improve the ecological situation (Fig. 14.1).

14.2.1 Immediate Vicinity of the Hydropower Plant; Significant Low Flow Velocities and Little Variation in Water Level

Possible measures:

- Creation of artificial or desilted existing oxbows with lateral connectivity to the river; mainly dredging work, creating raw sand areas and improvement of negative stagnant reed front, deep and shallow water conditions and adding temporarily change

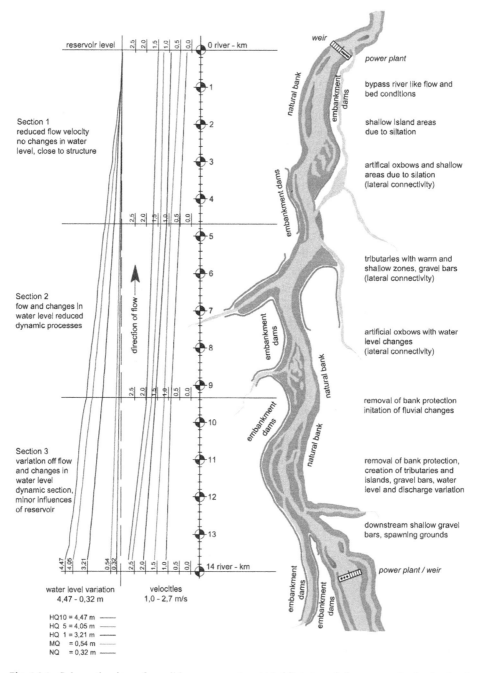

Fig. 14.1 Schematic view of possible measures to add habitats to existing reservoirs (water level variation (0–4.5 m; upper axis) and flow velocity (0.35–2.5 m/s; lower axis) for different design discharges (HQ10, HQ5, MHQ, MQ und NQ))

- Allow possible permanent flooding (connected from downstream, due to sand freight) of longitudinal areas that were formerly dry, sand removal from oxbows deposited during extreme floods
- Longitudinal connectivity; either technical solution in gorge like sites or if possible long bypass systems with gravel bed, flow variation, river like conditions in flow variation (water level).

14.2.2 Middle Part of the Reservoir; Minor Flow Variation and Temperate Seasonal Variation in Water Level

Possible measures:

- Desilt existing oxbows and allow lateral connectivity to the river; mainly dredging work similar Sect. 14.1
- Allow and add water level and floodplain variation especially during higher floods at shallow areas—different measures in the flood plains
- River shore and flow variation due to newly constructed structures: groins, gravel, stones from existing historic bank protection works and add shallow areas and terrestrial succession
- Connect and desilt tributaries to allow lateral connectivity and add gravel function.

14.2.3 Head of the Reservoir/Downstream Part of the Next Plant; High Flow Variation, High Seasonal Variation in Water Level and Therefore Almost Natural Variation (Water Level, Flow Conditions)

Possible measures:

- Artificial islands, shallow gravel bars (gravel function), inflow of bypass structure, add oxbows with lateral connectivity, add floodplain characteristic with deep and shallow water depth conditions and change of water level
- Remove shore protection, add river shore and flow variation with added structures; groins, gravel, trees, stones from existing historic bank protection works, add shallow areas during floods and allow terrestrial succession
- Connect and desilt tributaries to allow lateral connectivity and add gravel function.

The main focus of the implemented measures (Fig. 14.2) at the existing reservoirs was

reconnection of   gravel bars,   longitudinal connectivity juvenile habitats
oxbows, desiltation   shallow areas,   fluviatile flow   in different
lateral connectivity   spawning grounds   sediment variation   sections

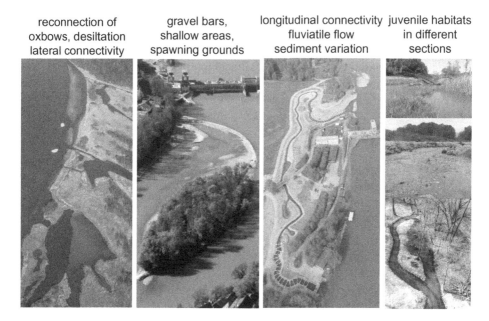

Fig. 14.2 All habitat components need to be available and reachable to meet the whole set of requirements of the life cycle. Nature like bypass channels can play a major role to add components to the anthropogenic changed river system

to allow and foster dynamic fluviatile changes and lateral connectivity to oxbows and tributaries. Especially the heads of the reservoirs often allow to meet the requirements of flow velocity and water level dynamics. Additionally, due to the historic river bed incision there are often minor restrictions in respect of flood protection. Therefore, the elimination of bank protection, the reconnection of existing and the creation of new floodplain areas and islands as well as the introduction of gravel bars into the system is possible. For these new alluvial structures, constructed from existing alluvial depositions, it is important to meet the aim of possible shallow gravel bars with shallow areas reaching up to high floods. Along the 250 km river Inn in Germany only around four to seven locations are yet identified to allow this sort of large scale restructuring. Restrictions are mainly flood protection risks, infrastructure, availability of land, settlements and influenced agricultural land. An important design criteria for the River Inn is to cope with the deposited sand in the flood plains and their natural deposition during and after floods. Therefore, an artificial furcation zone needs to be designed in respect of sheer stresses that only minor sand deposition occurs. Sand is part of the natural river system of the Inn and its transport and deposition is accepted and supported. For artificial backwater zones—an upstream connectivity—would result in a fill up within one season. The existing deposited sand has been used to create protection structures around the artificial oxbows or was added to the River Inn, as transport and deposition on land is almost impossible. The Inn system does transport the sand on a regular basis without any recognisable change or environmental impact.

More than seven years after the implementation of the first structures some of them still have the function as expected, others were dominated by hydrological events with its sand impact, some show regular natural variation in structure and habitat and some need regular maintenance to guarantee connectivity or spawning ground conditions. In a natural river system with its variation in hydrology and sediment transport some approaches targeting the aquatic zones became later terrestrial dominated. However, in an integrated approach the major factor and aim is, allowing changes within the system to happen. Sometimes the river dominates the system, sometimes the anthropogenic changes are so dominant that process oriented maintenance should be considered on a regular basis. But such an approach becomes difficult to implement if protected species might be affected, either by the maintenance measures or by the natural changes in the zoonosis.

14.3 Near Natural Bypass Systems as Key Element of an Integrated Approach

Upstream connectivity for fish migration is a main requirement to meet the WFD criteria in Europe. Guidelines have been developed mainly using the experience in salmon rivers and at small HPPs in (northern) Europe. Especially guidelines on the findability of the entrance such as flow velocity the "competing" discharge, and the location of entrance are difficult to be reasonably implemented at existing plants and local site conditions at large alpine rivers. For example, in the Inn the mean summer discharge varies from 500 to 1,000 m^3/s (downstream Salzach River confluence) and high turbulences and flow velocities can be observed downstream of the plant (up to 200 m). Therefore, the guidelines on entrance location can seldom be met. Existing guidelines only can provide a rough basis for the design at large rivers. Furthermore, the combined effects of connectivity and the creation of the habitat requirements of the fish species to the river system is recommendable and the following aspects and criteria should be fulfilled.

The main design criteria for bypass-systems on the Inn River are:

- Only potamodromous fish species (no salmon and no eel)
- All present fish species (around 40–50) shall reach the habitat seasonally to fulfil their life cycle
- All required habitats within the reservoirs should be available, reachable and should meet the quality requirements to fulfil the life cycle
- To reach spawning areas, add juvenile habitats, food habitats as well as flood and winter habitats are essential.

General layout for bypass channels and connectivity:

- Connection up and downstream (genetic cross over and compensation migration especially after floods); mainly juvenile and small fish, adult fish
- Lateral connectivity to adjoining tributaries, streams, oxbows and small pools—habitat variety for different age classes
- Spawning grounds with substrate variety, shallow areas typically for a natural river reach and typical habitat layout, including mainly juvenile habitat aspects
- If no constraints (adjoining owner, ownership, geometrical conditions) exist: variation of discharge, duration of flow, water levels, sediment transport—mainly river like design criteria slope, depth, velocity and sediment variation
- Detectability of entrance due to attractive structures, flow conditions at specific site and prioritisation of habitat aspects versus pure connectivity criteria.

Starting with the first sites in 2013 until now all experts involved in the layout and design as well as the experts at the authorities agreed in the local adaption of the guidelines design criteria to meet the overall aim—to increase and add a large number of habitats necessary—to the Inn River system. A variety of different types of bypass channels (size and discharge), technical structures and adaptions to the local site conditions were implemented. For experts without local knowledge it is often difficult to understand the local constraints (land use, ownership, geometrical constraints (gorges) and third party influence during the approval process, fears from flooding etc.) which restrict some originally intended design criteria. However, even small bypass channels with constant discharge can provide missing habitats for key fish species for all life-stages. Since 2015 all aspects of our bypass channels, habitat variations and instream structures are systematically evaluated by the Technical University Munich (Nagel et al. 2019). Spawning rates, numbers of juveniles, number of species and change of species, habitat use over time, are evaluated in the project lasting in total 10 years. In the following 5 years an individual tracking and tracing of more than 20.000 fish is commissioned to be installed and evaluated to get further information of travel distance, habitat use (bypass and special habitats) and the seasonal habitat use of the Inn fish population. In 2027 a connectivity of almost all plants from Switzerland to Vienna is reached.

References

Holzner M, Loy G, Schober HM, Schindlmayr R, Stein C (2014) Vorgehensweise zur Entwicklung von populationsunterstützenden Maßnahmen für die Fischarten am Inn in Oberbayern. Wasserwirtschaft Jhr 104(7/8):18–25
Loy G, Holzner M, Schober HM, Schindlmayr R, Stein C (2014) Maßnahmen zur Förderung von Populationen bedrohter Fischarten am Inn (Oberbayern) im Rahmen des Gewässerunterhaltes. Wasserwirtschaft 104(7):826–33

Nagel C, Müller M, Pander R, Geist J (2019) Bewertung von habitatverbessernden Maßnahmen zum Schutz von Fischpopulationen, Projektjahr 2019, unveröffentlicht. Lehrstuhl für aquatische Systembiologie, TU München

Reckendorfer W, Loy G, Ulrich J, Heiserer T, Carmignola G, Kraus C, Zemanek F, Schletterer M (2017) Maßnahmen zum Schutz der Fischpopulation—die Sicht der Betreiber großer Wasserkraftanlagen. WasserWirtschaft 2–3:82–86. https://www.springerprofessional.de/massnahmen-zumschutz-der-fischpopulation-die-sicht-der-betreibe/12114410

Risk Assessment and Decision Making on Mitigation Measures

15

Ruben van Treeck⊙, Christian Wolter⊙, Ian G. Cowx, Richard A. A. Noble⊙, Myron King, Michael van Zyll de Jong, and Johannes Radinger

15.1 Introduction

Sustainable development aims for decarbonized renewable energy to combat the impacts of climate change. Hydropower is one mode of renewable energy that actually contributes about 16% to the global gross electricity production (IEA 2020). Capturing energy from rivers is in line with the revised EU Renewable Energy Directive (Directive (EU) 2018/2001 of 11 December 2018), which established at least 32% share of

R. van Treeck (✉) · C. Wolter · J. Radinger
Leibniz Institute of Freshwater Ecology and Inland Fisheries, Berlin, Germany
e-mail: ruben.vantreeck@ifb-potsdam.de

C. Wolter
e-mail: christian.wolter@igb-berlin.de

J. Radinger
e-mail: johannes.radinger@igb-berlin.de

I. G. Cowx · R. A. A. Noble · M. King
Department of Biological and Marine Sciences, Hull International Fisheries Institute (HIFI), University of Hull, Hull, UK
e-mail: i.g.cowx@hull.ac.uk

R. A. A. Noble
e-mail: r.a.noble@hull.ac.uk

M. King
e-mail: mking@grenfell.mun.ca

M. van Zyll de Jong
Department of Biological Sciences, University of New Brunswick, Saint John, NB, Canada
e-mail: m.vanzylldejong@unb.ca

P. Rutschmann et al. (eds.), *Novel Developments for Sustainable Hydropower*,
https://doi.org/10.1007/978-3-030-99138-8_15

the Union's gross final consumption as new binding target for the EU by 2030, with a possible upwards revision already in 2023.

Directive 2018/2001/EU also advocates measures to support small renewable schemes through direct price initiatives such as feed-in tariffs (Article 17). Such incentives supported refurbishment and new installation of renewable energy schemes, including small-scale hydropower schemes in EU Member States. However, hydroelectricity generation causes a variety of environmental effects (compare Chap. 4) causing trade-offs between carbon emission-free energy generation and environmental impacts, which contradict other EU policies like the Water Framework Directive or the Biodiversity Strategy. Still, there is no commonly agreed, standardized and reproducible environmental risk and impact assessment of hydroelectricity generation to inform decisions on commissioning new or refurbished hydropower plants.

This chapter presents the first comparable environmental hazard scoring tool for risks and impacts of hydropower plants on fishes and further guidance for assessing cumulative effects of consecutive barriers and hydropower plants within a river to support informed decisions to mitigate environmental impacts from hydropower.

15.2   The European Fish Hazard Index

The trade-off between renewable hydroelectricity and environmental risks and impacts of hydropower plants (HPPs) on river ecosystems needs careful consideration at every single hydropower scheme. A suitable assessment framework evaluates hazards of one hydropower constellation relative to others as a function of their specific operational, constructional and technical characteristics and the ambient fish community, its sensitivity and species-specific mortality risk.

The European Fish Hazard Index (EFHI, van Treeck et al. 2021), is a fish-based assessment tool for screening the risks of hydropower for fishes and meets these requirements. It translates conceptual and empirical knowledge about hydropower-related hazards for fishes into site-specific risk scores. The EFHI is designed as modular assessment framework that offsets the relative hazard of a planned or existing HPP with the susceptibility of the local fish assemblage. It explicitly considers specific autecological characteristics of local species (van Treeck et al. 2020), their conservation value and specific regional management targets. The EFHI is applicable across all European biogeographic regions and can be adjusted to local environmental conditions and conservation objectives by selecting particular target species. At the same time, the EFHI is comprehensive and sufficiently versatile to be applied to a wide range of HPP designs in various stream types.

The EFHI integrates both, species-specific sensitivities of the fish community derived from species' life-history traits and conservation value as well as specific operational, constructional, and technical characteristics of an HPP. Principally the EFHI's hazard components and the final EFHI score increase with increasing severity of operational, constructional, and technical hazards of the HPP and with increasing sensitivity of the affected species community. The highest possible EFHI scores will be assigned to hydropower plants posing the highest overall mortality risks to fishes, installed in streams with numerous sensitive or conservation-critical species.

EFHI considers five constructional, technical and operational aspects of hydropower schemes that directly affect fishes: (i) upstream and downstream flow alterations, (ii) entrainment risk, (iii) turbine mortality, (iv) upstream fish passage, and (v) downstream fish passage (Fig. 15.1). These factors and their specific hazards are extensively described by van Treeck et al. (2021). Therefore, this chapter only briefly refers to the EFHI's mechanistic functioning. The EFHI considers hazards in relation to the characteristics of the ambient, site-specific fish community. This is captured by using the sensitivity of species to additional mortality (van Treeck et al. 2020) to weigh hazards.

The EFHI requires information about the: (i) HPP's main dimensions, turbine specifications, operating conditions, fish migration facilities and fish protection installed, (ii) target species, and (iii) river reach characteristics. Hazard thresholds were derived from conceptual and empirical knowledge or model results and subsequently categorized to

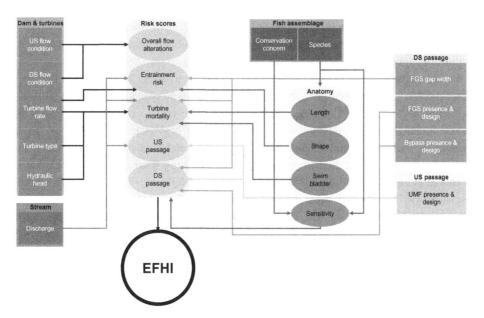

Fig. 15.1 Mechanistic model of the European Fish Hazard Index components. Rectangles = input parameters, ellipses = derived variables, open circle = final index. The "risk scores" box represents the process of transferring risk classes, species sensitivities and anatomies into adjusted and unadjusted hazard- and species-specific risk scores as shown in Table 15.1, which are aggregated to the EFHI. Figure obtained from van Treeck et al. (2021)

"high", "moderate" or "low" risk. These categories were cross-tabulated and weighed by the rounded integer value of the species' biological sensitivity (obtained from van Treeck et al. 2020) and yield a numerical score for each hazard and target species as shown in Table 15.1. Component- and species-specific hazard scores can take values from 0 to 1 in 0.25 steps, with higher scores indicating more severe hazards. Up to five target species can be selected to best reflect the local fish assemblage, conservation concerns and river reach. Target species can be manually assigned to the highest sensitivity class, regardless of their original score, to account for regional conservation concerns and management targets.

The single elements of the EFHI will be briefly described in the following.

15.2.1 Flow Alterations

The EFHI assesses the impact of both upstream and downstream flow alterations. Upstream flow alteration is assessed based on the HPPs' reservoir storage capacity, which is typically given relative to the average net inflow per time period (Langbein 1959, McMahon and Mein 1978). Reservoirs with a storage capacity exceeding the average annual net inflow are scored high risk. Smaller reservoirs and impoundments were further discriminated by their average flow velocity: Those reducing the mean flow velocity below 0.5 m/s were classified moderate risk and those still maintaining mean flow velocities of 0.5 m/s or higher low risk.

Downstream flow alterations are typically attributed to either hydropeaking or water abstraction (e.g., in residual flow stretches of diversion-type plants). Because hydropeaking inherently results in a completely altered discharge regime with severe impacts on stream biota, this operation mode was always scored high risk. The hazard of water abstraction, particularly problematic in residual river stretches of diversion schemes, was scored by environmental flow recommendations following Tharme (2003). Remaining discharge below 10% of the mean annual low flow (MNQ) was scored high risk. Higher discharge, but less than or equal 10% of mean annual flow (MQ) was scored moderate risk and >10% MQ low risk. The overall downstream risk was assigned according to the higher risk class. Upstream and downstream hazard classes were aggregated into the total flow alteration hazard using the same principle.

Table 15.1 Weighed numerical impact scores for specific HPP hazards and sensitive species for calculating of the EFHI. Target species sensitivity scores from low (2) to high (4) from van Treeck et al. (2020)

	Target species sensitivity		
Hazard classification	4 (high)	3 (moderate)	2 (low)
High	1	0.75	0.5
Moderate	0.75	0.5	0.25
Low	0.5	0.25	0

15.2.2 Entrainment and Turbine Mortality (ETM)

The total flow rate of all installed turbines was used as proxy of the entrainment risk for fishes describing the probability of fishes passing through the turbines rather than taking any other route downstream. This risk was estimated as ratio of flow rate to mean discharge (MQ) and scored high at ratios ≥ 1, moderate between >0.5 and <1 and low ≤ 0.5.

To account for the fish-deflecting effect of installed fish guidance structures (FGS), the entrainment risk score was adjusted for each installed turbine and turbine mortality hazard score as follows: Width-to length-ratios of target species empirically derived by Ebel (2013) to estimate the maximum length of both eel-like and non-eel-like (e.g., fusiform) species that could physically pass the FGS at a given bar spacing.

The turbine-specific mortality risk was assessed based on three components: (i) model-based turbine blade strike rates for Francis and Kaplan type turbines, (ii) empirical mortality rates for Archimedic screws, Kaplan very-low-head, Ossberger, Pelton, Pentair Fairbanks and water wheels, and (iii) empirical turbine barotrauma mortality rates. Following Wolter et al. (2020), mortality rates between 0 and 4% we scored low, between $>4\%$ and 8% moderate and $>8\%$ high risk. To evaluate HPPs with more than one turbine, the EFHI assigns risk scores for each individual turbine and species as shown in Table 15.1 and, subsequently, aggregates the relative contribution of each turbine to the overall mortality rate weighted by the turbines' specific flow rates. The turbine with the highest flow rate gets the proportional relevance "1" and all further turbines an equal or proportionally lower proportional relevance depending on their flow rate. The proportional relevance of each turbine serves to weigh the mortality risk score.

We applied the frequently used blade-strike model by Montén (1985) that calculates the probability of fishes striking a blade depending (among others) on their length and the space between turbine blades. For these models we used either the body length of fishes being physically able to pass the FGS or common length of an adult specimen if no FGS was installed. Modelled mortality rates for Francis and Kaplan turbines were classified low (0–4%), moderate (>4–8%) and high (>8%), respectively.

Other turbines without available mortality models were scored based on empirical data. Low risk turbines were water wheels (<1% mortality, Schomaker and Wolter 2016), Pentair Fairbanks (<1% mortality, Van Esch and van Berkel 2015; Winter et al. 2012), and very-low-head turbines (VLH, <4%, Hogan et al. 2014; Lagarrigue et al. 2008; Fraser et al. 2007). Of moderate risk were Archimedes screws (<8%, Wolter et al. 2020) and of high risk Ossberger (>99%, Gloss and Wahl 1983, Knapp et al. 1982) and Pelton turbines (>99%, Cada 2001).

Modelled strike mortality rates were complemented with a risk assessment of pressure-related injuries derived from the HPP's hydraulic head. Barrier height < 2 m was scored low risk, 2–10 m moderate and >10 m high risk based on empirical data (Wolter et al. 2020). The hydraulic head risk score was further adjusted according to species-specific susceptibility to barotrauma and quantitative model observations by Wilkes et al. (2018).

The risk score was set to zero for species without swim bladder and thus, not experiencing barotrauma risk. The risk score remained unchanged for physostomous fishes with an open swim bladder allowing for quicker pressure compensation, and it was raised by 50% for physoclistous fishes with closed swim bladder unable to quickly balance pressure changes (Brown et al. 2013, 2014; Colotelo et al. 2012; Harvey 1963; Wilkes et al. 2018).

15.2.3 Upstream Fish Passage

A major driver of the effectiveness of upstream fish passage facilities is their discharge relative to the mean discharge of the river, with higher values increasing passage success (Wolter and Schomaker 2019). We used two linear regressions to determine the minimum recommended discharge in an upstream migration facility as a function of the discharge of the stream. One was applied to rivers with a mean discharge up to 25 m^3/s and a discharge in the upstream migration facility between 3 and 5% of that value, while the other was used for larger rivers that need less proportional discharge in the upstream migration facility (between 1 and 3%). Input discharge values as well as calculated "best-practice" values were rounded to one decimal point and compared. If the actual discharge in the upstream migration facility equalled or exceeded recommendations, the risk class was scored low. A discharge $\geq 50\%$ of the calculated recommendation was scored moderate risk. Discharges <50% of the recommendations or no upstream migration facility were both considered inadequate and scored high risk. The superior passage performance of nature-like fishways was acknowledged by lowering both the upstream and downstream passage score by 20% each (in combination with FGS preventing turbine passage).

15.2.4 Downstream Fish Passage

Hazards for downstream migrating fish were scored as follows: Angled bar racks, louvers, modified and curved bar racks at a horizontal installation angle of $\leq 45°$, or vertically inclined bar racks with an inclination of $\leq 45°$, all in combination with downstream bypasses accessible across the whole water column have proven efficient (Calles et al. 2013a, b; Ebel 2013) and scored low risk. FGS installed at larger horizontal or rather steep vertical angles are demonstrably less efficient. Although they still prevent fishes from entrainment, they also cause damages or mortality due to impingement or shear forces (e.g., Calles et al. 2013a, b, 2010; Larinier and Travade 1999). In addition, the efficiency of bypasses accessible only at discrete location in the water column is highly variable (Calles et al. 2013a, b; Gosset et al. 2005; Økland et al. 2019; Travade et al. 2010). Therefore, vertically inclined bar racks and horizontal FGS with angles >45° as well as any constellation without fully accessible bypass were scored as moderate risk.

A missing FGS or bypass was scored as high risk. Similar to upstream fish passage hazards, the bi-directional performance of a nature-like fishway is rewarded by reducing the downstream passage hazard score by 20%.

15.2.5 Final EFHI Score

The calculations outlined above produce in total 20 single score values: four hazard components individually assessed for five fish species each. Their aggregation is conducted in two steps; first the arithmetic mean is calculated for each component and second, the component means are averaged to the final EFHI score. Final scores ≤ 0.33 were classified "low risk", scores between >0.33 and 0.66 were classified "moderate risk" and scores >0.66 were classified "high risk".

15.2.6 Application of EFHI

To demonstrate the EFHI, it was applied to a small hydropower installation (<1 MW) in southern Germany. The plant creates only a small impoundment, but the flow speed therein is relatively slow (<0.5 m/s). The scheme is a diversion plant, with a well water-supplied residual river stretch. It has one Kaplan turbine and two Archimedes screws installed. The Kaplan turbine is four-bladed with 2.5 m outer and 0.95 m hub diameter and 100 rounds per minute (RPM) rotational speed. The total flow rate of all three turbines is 36 m^3/s, of which 18 m^3/s go through the Kaplan turbine and 9 m^3/s each through the two screws. The Kaplan runner is equipped with a vertical bar rack with 20 mm bar spacing and the two screws are protected by a 150 mm trash rack. The upstream migration facility has 1 m^3/s admission flow. A downstream bypass is not installed. The surrounding fish fauna is dominated by rheophilic cyprinids; the target species used in EFHI are chub *Squalius cephalus* (sensitivity 3.0), barbel *Barbus barbus* (4.1), minnow *Phoxinus phoxinus* (2.5), spirlin *Alburnoides bipunctatus* (2.1) and dace *Leuciscus leuciscus* (3.1). Applying the EFHI calculation software tool (available at https://zenodo.org/record/468 6531) as briefly described above, this particular constellation of HPP and river characteristics, target fish species and conservation concerns yields an overall EFHI score of 0.51 indicating a moderate risk.

To enhance fish protection, the plant could be refurbished with a finer screen of 15 mm bar spacing only and a fully accessible downstream bypass. In addition, increasing the flow speed in the impoundment to >0.5 m/s would further improve the EFHI score. These measures together would reduce the overall EFHI score to 0.31. In contrast, if there would be no functioning upstream fish migration facility and the bar spacing of the fine screen widened to e.g., 30 mm, the same plant would be scored 0.58. These example calculations demonstrate the EFHI application to assess existing hydropower plants and the effects of potential mitigation measures and modifications of plant components on the hazard risk for fishes.

The EFHI can also be used to compare risk variations of similar HPPs in response to differences in sensitivity and conservation value of the ambient fish assemblages. For example, if the same HPP characterized above is located in an European eel (*Anguilla anguilla*) catchment hosting more sensitive fish species (e.g., trout *Salmo trutta* and nase *Chondrostoma nasus* instead of minnow and spirlin), the higher conservation value of the fish assemblage would substantially raise the final EFHI score to 0.68 "high risk", even though stream size and HPP components remain unchanged.

The hazard components and final EFHI scores of the four scenarios outlined above are displayed in Fig. 15.2.

15.3 Cumulative Impact Assessment

The larger the rivers are the more they become exposed to multiple pressures, which might antagonistically or synergistically interact and result in cumulative impacts of pressures. While a variety of pressures might create cumulative impacts, the following section addresses only cumulative impacts of a series of barriers with or without hydropower plants in a river. It aims to provide guidance that allows hydropower developers as well as water managers to consider effects at a catchment scale, and thus maintain a high level

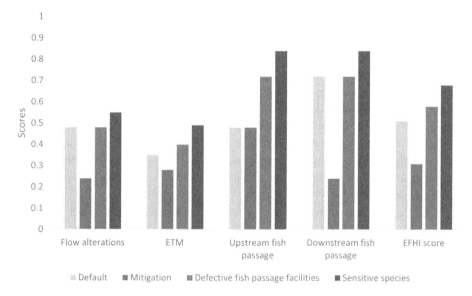

Fig. 15.2 EFHI scores of an exemplary small hydropower plant in default operation mode (yellow), after improving fish safety (blue), after failure of the fish protection measures (brown) and assuming a more sensitive ambient species community (purple)

of environmental protection in line with the Water Framework and Habitats Directives. The cumulative impact assessment (CIA) presented here provides valuable inputs that need to be considered when a hydropower scheme is proposed and developed for construction or retro-fitted to mitigate potential impacts. However, it is important to recognize that generic models cannot be built for this purpose, because every scheme is different and its contributions to power production will vary depending on river conditions and operational practices. Whilst offering generic guidelines for assessment of cumulative impact of multiple hydropower schemes, CIA is limited in its assessment of risks and uncertainty about the impact of individual schemes on fisheries and the environment, which must be assessed using scheme-specific impact assessments and then used to understand the role of individual schemes in a wider cumulative assessment. The CIA protocol is constructed to support the Decision Support System for hydropower schemes that is presented in Sect. 15.4 of this chapter.

The CIA approach considers all transversal barriers in a river system. Depending on available migration facilities for fish, barriers form more or less significant migration obstacles for fish, fragment habitats and populations. Barriers are also obstacles for flow and sediment transport creating impoundments upstream with lower flow velocities and higher sedimentation, which typically result in habitat loss for rheophilic, gravel spawning river fishes. Therefore, the CIA approach also considers the cumulative length of impoundments in a river system, which corresponds to habitat loss for river fishes. It must be noted that numerous eurytopic and stagnant water-preferring species will benefit from impoundments. However, this benefit is not positively counted, because it does not correspond to the type-specific fish community and thus to a good ecological status of the river system according to the Water Framework Directive. Finally, some of the barriers are used for hydroelectricity production. Here in addition to the other impacts also fish mortality at hydropower plants might accumulate and is considered by the CIA. The different spatial elements considered in the CIA approach are illustrated in Fig. 15.3.

From a fish-ecological perspective it has to be considered, that diadromous species—these are species that obligatorily use marine and freshwater habitats during their life cycle—essentially have to pass all obstacles between their marine and freshwater habitats. Therefore, these species, e.g. European eel and Atlantic salmon, will experience the full cumulative effects of all pressures. Potamodromous species are also obligatory migrants, but they complete their life cycle only in freshwaters. Depending on habitat availability, they might pass only a limited number of barriers that will exert cumulative impacts.

The majority of non-obligatory or facultative migrating species does not depend on regular migrations and might complete their life cycle even within river fragments depending on habitat availability. However, all fishes show more or less extended exploratory movements and homing behaviour, with larger fishes moving longer distances and using larger home ranges (Radinger and Wolter 2014). Therefore, in particular the large-bodied species face a higher encounter probability with barriers and hydropower plants, but they hardly have to pass more than one barrier and thus will be least impacted by cumulative effects.

Fig. 15.3 Elements of
cumulative impact assessment:
cumulative share of
impoundments/habitat loss for
river fishes, cumulative number
of barri-
ers/fragmentation/migration
obstacles, and cumulative
mortality at hydropower plants

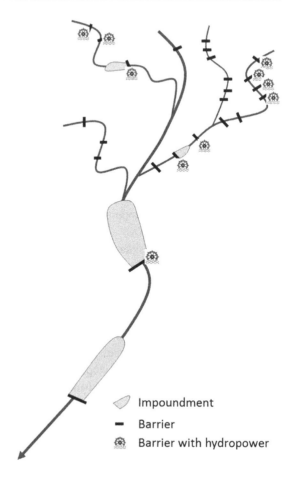

◁ Impoundment
▬ Barrier
🌀 Barrier with hydropower

Independent of their migratory life history trait, all fish species will be affected by the cumulative length of impoundments that changes the hydromorphologic river character and by the fragmentation of habitats that increases with the number of barriers in a river system.

15.3.1 Cascades of Consecutive Impoundments

Generally, the natural power potential of Europe's watercourses is heavily utilized for energy production, particularly in alpine areas such as Norway, Austria, Switzerland or Southern Germany where topographical and hydrological conditions are most suitable. For example, the degree of hydropower development is already at a level of about 95, 90 and 70% in Switzerland, the Federal province Upper Austria and Bavaria, respectively (Jungwirth et al. 2003). At a global level, conservative estimates assume a total surface area of reservoirs of more than 400,000 km² (source: internationalrivers.org), which does not include habitat losses generated by smaller dams and weirs. As a result, 15% of the global annual river runoff is stored in reservoirs (Likens 2010) and 48% of the global river volume is moderately to severely impacted by flow regulation, fragmentation or both (Schmutz and Moog 2018). Moreover, hydropower plants often occur in cascades creating a series of impounding reservoirs in one stream and thereby cumulatively contributing to the overall change of the hydraulic and habitat conditions of otherwise naturally free-flowing ecosystems.

Therefore, it is a main objective to quantify and assess the habitat loss caused by the cumulative impoundment effects of multiple anthropogenic barriers (i.e. cascades of dams and weirs) that alter the natural flow and habitat conditions of rivers with impacts on the riverine fish community. This involves several methodological questions related to the estimation/quantification of the spatial effects of impoundments (e.g. how large are impoundments and how much habitat area is altered) as outlined below.

Delineating Impoundments

A rather obvious way to delineate river impoundments is their direct on-site measurement in the field. While this might be considered time and resource consuming, it is very accurate to map flow modifications on-site, i.e. to assess deviation from natural/reference flow conditions and to determine the spatial extent of the impoundment in an upstream direction, especially in smaller streams. For example, to facilitate the assessment of flow modifications in Germany, the State Agency for Nature, Environment and Consumer Protection (LANUV) provided an overview of reference conditions of naturally flowing rivers based on hydromorphological river typologies (Timm et al. 1999).

Alternatively, the impoundment effect can also be roughly calculated from the dam height and the natural stream slope. Therefore, the dam height, HD is simply divided by the bottom slope S:

$$\Delta x = HD/S$$

For example, a 1 m high dam in a stream with a natural slope of 2 ‰ (0.002) creates and impoundment of about 500 m length.

Impact Assessment of Habitat Loss Due to Impoundments

We consider the total sum of lengths of all impounded sections of a river relative to the remaining free-flowing sections as the most decisive measure of the cumulative effect size of impoundments. Therefore, the cumulative length of impoundments per river section is divided by the total length of that river section. As a section, we consider the segment of the river between two confluences, i.e. points where two rivers merge. This follows the general principle of portioning streams into segments, which is also used in classical stream order concepts. This assures that relatively small impoundments of smaller streams are not compared to the entire (potentially free-flowing) main stem which generally provides habitat for different fish communities. Vice versa, habitat losses due to impoundments in the main stem (or higher order rivers) should be addressed independent of (potentially free-flowing) head water sections, to account for the fact that different parts of the river along the longitudinal gradient are inhabited by different fish communities.

To provide an example, we followed the outlined approach to calculate the relative length of all impoundments per river section (= cumulative length of impoundments of river section i/total length of river section i) for all Austrian rivers. Similarly, we calculated the relative barrier density (= number of barriers of a river section i/total length of river section I). Spatial data on impoundments, barriers and the watercourse were obtained from the Water Information System Austria (WISA, https://maps.wisa.bmlrt.gv. at/gewaesserbewirtschaftungsplan-2015). Vector maps indicate in-situ mapped impoundments. River sections were extracted as lines between two confluences. For the visual analysis relative share of impoundments per river section as well as barrier density data were grouped into classes of stream order ranging from headwater streams (classes 1–3) to large lowland rivers (class 9, River Danube). The results show that the relative cumulative share/length of impoundments is increasing with increasing stream order (Fig. 15.4) while the barrier density is highest in low order streams and decreasing downstream. This becomes especially evident for the River Danube (stream order = 9) where most river sections are already impounded and only a few free-flowing river sections remained, whereas the number of barriers (i.e. barrier density) is comparably low. By comparison, sections of lower order head water streams might have higher numbers of small weirs and thus many impounded sections; however, these impoundments are usually rather short and there are still many non-impounded river sections.

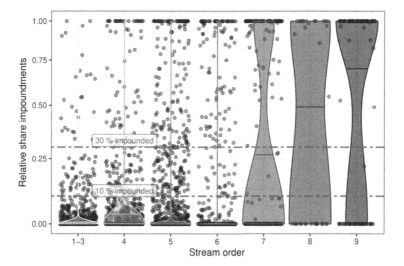

Fig. 15.4 Relative cumulative length of impoundments per river section and across stream orders in Austria (data from Water Information System Austria). Left to right is the upstream–downstream gradient. Points show relative cumulative impoundment length for single river sections. Violin plots indicate density distribution for a given stream order class

15.3.2 Fragmentation Due to Multiple Barriers

In addition to the alteration of river habitats due to impoundments, hydropower and their associated dams and weirs fragment river networks. Barriers represent one of the largest anthropogenic impacts on river ecosystems and limit habitat connectivity at multiple spatial and temporal scales (Fuller et al. 2015). This might express in particular in the impediment of ontogenetic migrations (e.g. spawning runs) and ordinary habitat movements of river fish (e.g. Marschall et al. 2011; Radinger and Wolter 2015; Radinger et al. 2018) and associated genetic fragmentation of populations (e.g. Gouskov et al. 2016). Recent studies emphasized that the location of a barrier within a river system and especially its location relative to suitable habitats and species occurrences determines its impact on fish (Kuemmerlen et al. 2016, Radinger and Wolter, 2015). Nevertheless, even moderate densities of barriers might not be acceptable given the commonly high mobility of river fish (Radinger and Wolter 2014). For example, Radinger et al. (2015) modelled the occurrence patterns of riverine fish in response to hydromorphological variables and in-stream structures and gained best modelling results when the conditions in a distance of 1–4 kms up- and downstream of a site were considered, thus indicating potentially extensive movements of fish within and between habitats.

Corresponding to previous studies (e.g. Van Looy et al. 2014), we consider the number of barriers per river km (i.e. relative barrier density) a suitable and easily obtainable indicator of the degree of fragmentation of a given river section. Analogous to the calculation of the relative share of impoundments, we consider a river section as the segment of the river between two confluences, i.e. points where two rivers merge. This assures that parts of the river network with rather low densities of barriers, as common in low-gradient streams and rivers, are not mixed with highly fragmented river network sections, as more common in high gradient streams as illustrated in Fig. 15.5 for Austrian rivers.

Assessment of Habitat Loss and Fragmentation Due to Multiple Cumulative Barriers

Based on the median values (half of the river sections per stream order), the cumulative length of impoundments was rather low in low order river sections and increased downstream (Fig. 15.4). This relation very well reflects the hydromorphic conditions of predominantly Alpine and Prealpine river systems, with steep slopes resulting in low impoundment length. However, the cumulative length of impoundments is a metric for habitat loss only and does not allow conclusions on the severity of habitat fragmentation and migration barriers. Inversely to the cumulative length of impoundments, the relative number of barriers per river kilometre was high in low order river sections and decreased in the downstream direction (Fig. 15.5).

These findings lead to a scoring of impacts and habitat loss resulting from impoundments and fragmentation due to barriers as shown in Table 15.2.

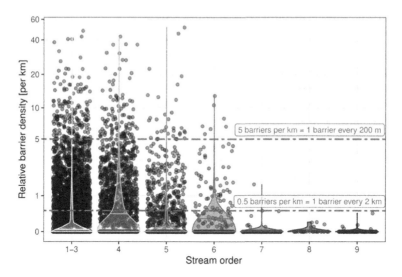

Fig. 15.5 Relative barrier density (number of barriers per km) per river section and across stream orders in Austria (data from Water Information System Austria). Left to right is the upstream–downstream gradient. Single points show relative barrier density for a single river section. Violin plots indicate the density distribution for a given stream order class

Table 15.2 Impact scoring of habitat loss due to relative cumulative impoundment lengths and barrier density

	Low impact	Moderate impact	High impact
% impounded length (relative to the total length of the river section)	≤10%	>10–30%	>30%
Relative barrier density (barriers per km, per river section)	<0.5 barriers/km	0.5–5 barriers/km	>5 barriers/km

Given that the highest hydromorphic and ecologic state of a river is free flowing with functioning sediment generation, transport and sorting processes and a nearly undisturbed riverine species community, a slight deviation might be 10% habitat loss. Therefore, habitat losses up to 10% in total might be considered of low cumulative impact. Accordingly, habitat losses of up to 30% in total in a river segment might be accepted as moderate cumulative impact, while higher losses of habitats have a high impact on riverine species communities. The high cumulative impacts from habitat losses in most segments in higher order rivers corresponds well with the reported habitat degradation and deviation from good ecological status according to the WFD.

Given the high mobility of riverine fish, a barrier every two kilometres or less (i.e. 0.5 relative barrier density) is considered moderate fragmentation impact. Barrier densities of five (=1 barrier every 200 m) clearly conflicts with the movement behaviour and home range of most river fish (Radinger and Wolter 2014) and thus is considered a high fragmentation impact.

15.3.3 Cumulative Impacts—Diadromous Species

Understanding and evaluating cumulative impact of multiple HPPs in a catchment context requires fitting the life history characteristics to a life cycle model of the target species. These vary for different migratory species groups and account for the impact of individual schemes on components of the population life cycle.

Conceptual frameworks to assess the cumulative impact of multiple hydropower schemes are provided in Figs. 15.6 and 15.7 for anadromous and catadromous species, respectively. Each framework provides a starting point for assessment based on a count of the original adult population and follow this through the life cycle determining where the population will be impacted and by what proportion in addition to natural mortality. The anadromous framework starts with upstream migration passed the scheme (upstream barrier effect), opportunity for spawning and recruitment to the extant population based on habitat availability and finally losses caused by downstream migration through the successive hydropower schemes (Fig. 15.6). The catadromous framework (Fig. 15.7) starts with adults departing the river habitats and accounts for losses caused by downstream

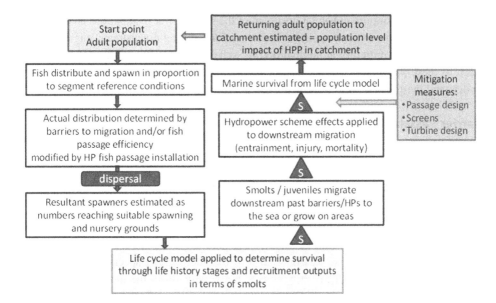

Fig. 15.6 Cumulative impact assessment for anadromous species—inputs and losses

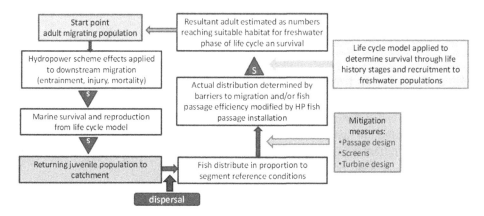

Fig. 15.7 Cumulative impact assessment for catadromous species—inputs and losses

migration through the successive hydropower schemes and then the impact on returning juvenile fishes dispersing in the target catchment following recruitment in the marine environment.

Each component of the life cycle can be modelled based on criteria specifically related to each individual scheme or based on field surveys. Where such data are not available, expert judgement or available information in the literature should be used to populate the models. Key elements in these frameworks are assessing impediment to upstream migration, loss of important habitat upstream of the hydropower installations and thus loss of recruitment to the extant population and loss of fish during downstream migration as a result of injury or mortality at the hydropower schemes or impacts of delayed migration. Each of these parameters can be quantified and the additive effect can be determined to understand the population impact of the cascade of hydropower schemes.

Barrier Passability

One of the key impacts of hydropower schemes is the disruption to connectivity caused by the dam structure. All hydropower schemes create a barrier to hold back or divert water to the turbine(s). The size of the barrier is highly variable depending on the design of the scheme but needs to be accounted for. An example of the cumulative impact of multiple barriers on a system is shown in Fig. 15.8. Here the impacts of seven barriers in succession on the population size of an upstream migrating species are compared with different levels of fish passability. It can be clearly seen that the cumulative effects of compromised passabilities < 0.5 (i.e. less than 50% of the approaching specimens successfully passed) at the barriers result in extirpation of the population in the upstream areas. It is thus essential to model the impact of variable passabilities at barriers to determine the cumulative impact.

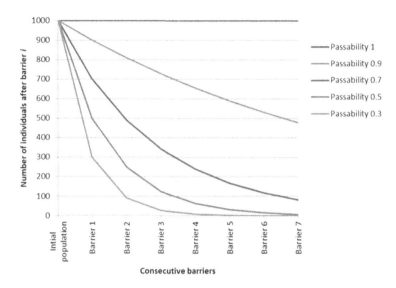

Fig. 15.8 Cumulative impact of multiple (seven) consecutive barriers of a given passability rate (from 0.3 = 30% of approaching fish to 1 = 100%)

A number of tools are available to assess barrier passability for fish (Kemp and O'Hanley 2010). A coarse resolution rapid barrier assessment methodology that is suitable for multiple fish species and considers both up and downstream dispersal was devised by Kemp et al. (2005) and later in 2012, by the Scotland and Northern Ireland Forum for Environment Research (SNIFFER). The assessment method uses rule-based criteria for fish morphology, behaviour, swimming and leaping ability to estimate barrier passability. The latter is the fraction of fish (in the range 0–1) that are able to successfully pass a given barrier. Each barrier is assigned one of four passability levels as follows: 0 is a complete barrier to movement; 0.3 is a high impact partial barrier, passable to a small proportion of fish or passable only for short periods of time; 0.6 is a low impact partial barrier, passable to a high proportion of fish or for long periods of time; and 1 is a fully passable structure.

Downstream Migration

Once fish have spawned, adults of many species must migrate downstream, either to the sea (anadromous species) or the lower reaches of the river (potamodromous species). The juveniles of both migratory guilds will also ultimately migrate downstream to complete their life cycles. In addition, adult catadromous species such as eel must eventually migrate downstream to complete their life cycles. In European rivers downstream migrating fish mostly actively swim downstream and may pass the dam by one of several corridors: downstream migration facility, turbine, water release over spillway or through sluice gates. In addition, it should be recognised that some juvenile life stages of fish move downstream by drifting in the current. This mode of migration past hydropower structures is dependent of the allocation of flows and with the majority water going through the turbines, most juveniles will experience the trauma of high pressure and sheer and will unlikely survive the experience.

As indicated, one route for downstream fish migration is passage through the turbine at hydropower installations, which can be related to injury and mortality, caused by several damage mechanisms as already outlined for the EFHI and in Chap. 4. Harrison et al. (2019) posit that overall passage efficiency for diadromous fish populations using turbine routes to pass hydropower dams in a downstream direction (DS Passage efficiency) can be considered a product of the conditional probability of reservoir, forebay and turbine entry, along with turbine passage survival, turbine passage exit survival, delayed survival and sub-lethal effects. These effects are again cumulative and result in an almost complete loss if downstream passage is low and mortality at the power station is high (Fig. 15.9), which is often the case.

15.3.4 Cumulative Impacts—Potamodromous Species

Addressing cumulative impact for potamodromous species follows a similar approach to diadromous species except that the species of concern only migrate within the river system, either up and downstream including into tributaries or onto floodplain systems to complete their life cycles. The schematic framework of cumulative impact assessment for potamodromous species is shown in Fig. 15.10 and the models can be populated based on information provided in the following sections on successive dam passage, home range and mobility and migration distances.

Successive Dam Passage
Few studies have investigated the passage success of non-diadromous species over more than one consecutive barrier in either an upstream or downstream direction.

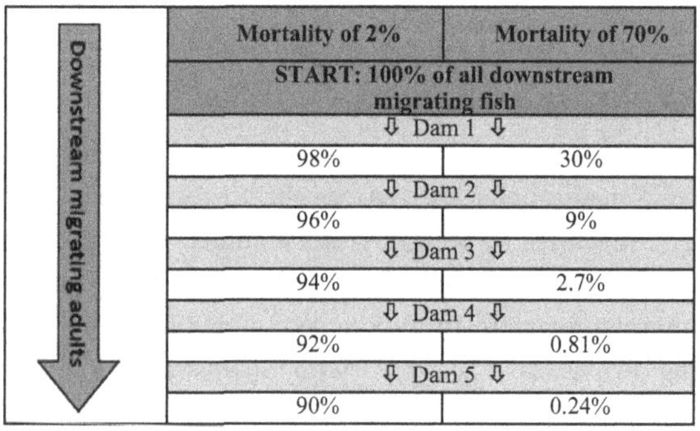

Fig. 15.9 Scheme for cumulative downstream passage efficiency for different mortality rates

Downstream migrating adults	Mortality of 2%	Mortality of 70%
	START: 100% of all downstream migrating fish	
	⇓ Dam 1 ⇓	
	98%	30%
	⇓ Dam 2 ⇓	
	96%	9%
	⇓ Dam 3 ⇓	
	94%	2.7%
	⇓ Dam 4 ⇓	
	92%	0.81%
	⇓ Dam 5 ⇓	
	90%	0.24%

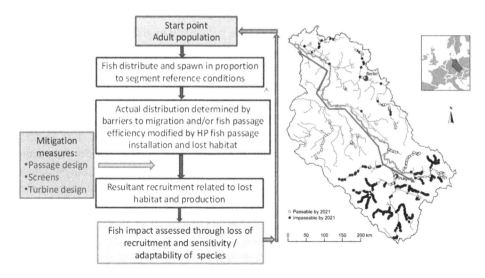

Fig. 15.10 Cumulative impact assessment for potamodromous species—inputs and losses

In the Dutch lower Rivers Rhine and Meuse, De Leeuw and Winter (2008) studied the movement of potamodromous fish species using transponders and a network of 29 detection stations. In total, 110 ide, 76 barbel, 51 chub and 8 nase were tagged, of them 57, 30, 15 and 4 specimens, respectively, were never detected during the study period between February 2003 and June 2006. Passage over weirs was observed for very few individuals only, as follows: of 2 approaching barbel 2 successfully passed the weir in an upstream direction, as did 3 of 4 chub, 1 of 2 nase, and 6 of 30 ide. Passage success in downstream direction was much less: 1 of 7 approaching barbel, 1 of 1 nase, 3 of 11 ide and not a single chub (De Leeuw and Winter 2008). The movement and passage over consecutive weirs of these four potamodromous species has also been studied by Benitez et al. (2018) in the Rivers Meuse and Ourthe in Belgium and by Lucas and Frear (1997) in the River Nidd (UK). All studies were performed using active or passive telemetry techniques and observed an average cumulative passage success of these species of 41% at the first barrier and 10% at the second.

Based on the little data available, there is evidence that migration of potamodromous cyprinids may fall to as low as 10% after bypassing two weirs. It has to be noted; however, that there might be less individual motivation in potamodromous fish for migrating larger distances compared with diadromous species, which also strongly depends on habitat quality and availability in the river reach between barriers (Benitez et al. 2018). It still remains challenging to identify the number of mobile specimens willing to pass longer distances over several barriers. More research is needed to gather information.

Migration Distance

To understand how far potamodromous species potentially migrate within river systems, empirical data on maximum migration distances have been compiled. They might serve as proxy to assess the maximum number of barriers a potamodromous fish has to pass during its life to complete its life cycle. The usual home range is typically much shorter. These migration distances vary according to river length and do not indicate long- or short-distance migrating as a kind of ecological trait of a species. Migrations have to cover the distance to the—not necessarily nearest (compare Fredrich et al. 2003, for chub *Squalius cephalus*)—spawning site or seasonal habitat. The spatial arrangement of river segments, river reaches and habitats strongly depends on river size and length. Accordingly, migration distances correlate with river size measured for example as Strahler of Shreve order. In larger rivers, the same species moves over significantly longer distances; therefore, river size and study time emerged as significant predictors of movement distance (Radinger and Wolter 2014).

Migration distances vary considerably within species depending upon individual motivation (Table 15.3). Fish personality, as well as genetic factors, seasonality, and purpose of the migration influence the distances covered (Brodersen et al. 2008; Brönmark et al. 2013). Resource migrations and diurnal and seasonal habitat shifts largely depend on the home range size (Fredrich 2003) and are thus related to river size (Radinger and Wolter 2014).

Also, discharge was reported to affect the annual migration distances as well as the intensity of spawning runs based on catch data for European sturgeons in the River Rhine (Kinzelbach 1987) and Atlantic salmon in the River Elbe (Wolter 2015). Especially in years with high floods, salmon were reportedly caught much further upstream than commonly observed (Wolter 2015). Similarly, further migration distances and improved passage of barriers must be assumed for potamodromous fish at higher discharges.

All potamodromous species are iteroparous, i.e. spawn several times during their reproductive life, which means that repeated upstream and downstream migrations for spawning are the rule. Typically, the downstream migration of the adults after spawning has no discrete runs or migration peaks but spreads over a longer period. Examples of downstream migration distances for potamodromous species are provided in Table 15.4.

Downstream migrations also take place when fish change sites to reach more productive or more suitable feeding grounds. Displacement through high water conditions or while searching for food may result in large scale downstream displacement. The same mechanism is effective during drift of larval fish or the undirected outmigration of juveniles. This movement is characterized by a non-uniform orientation towards the current which largely contrasts with active movement with the flow that is observed during adverse conditions (Lucas and Baras 2001).

Table 15.3 Empirical upstream migration distances of potamodromous and facultative migrating species

Species	Distance (km)	River	Method	Reference
Potamodromous migrants				
Aspius aspius	166	Elbe	Telemetry	Fredrich (2003)
Barbus barbus	22.7	Ourthe, Belgium	Telemetry	Ovidio et al. (2007)
Barbus barbus	12.7	Ourthe, Belgium	Telemetry	Ovidio et al. (2007)
Barbus barbus	>25	Meuse, Belgium	Telemetry	De Leeuw and Winter (2008)
Barbus barbus	2–20	Nidd/Ouse, England	Telemetry	Lucas and Batley (1996)
Barbus barbus	20	Nidd, England	Telemetry	Lucas and Frear (1997)
barbus barbus	34	Severn, England	Mark-recapture	Hunt and Jones (1974)
Barbus barbus	318	Danube	Mark-recapture	Steinmann et al. (1937)
Barbus barbus	2	Jihlava, Czech Republic	Mark-recapture	Penaz et al. (2002)
Chondrostoma nasus	>25	Meuse, Belgium	Telemetry	De Leeuw and Winter (2008)
Chondrostoma nasus	140	Danube	Mark-recapture	Steinmann et al. (1937)
Leuciscus cephalus	169	Danube	Mark-recapture	Steinmann et al. (1937)
Leuciscus cephalus	>25	Meuse, Belgium	Telemetry	De Leeuw and Winter (2008)
Leuciscus cephalus	25	Spree	Telemetry	Fredrich et al. (2003)
Leuciscus idus	>45	Meuse, Belgium	Telemetry	De Leeuw and Winter (2008)
Leuciscus idus	19	Elbe	Telemetry	Kulíšková et al. (2009)
Leuciscus idus	>100–187	Elbe	Telemetry	Winter and Fredrich (2003)
Leuciscus idus	82	Danube	Mark-recapture	Steinmann et al. (1937)

(continued)

Table 15.3 (continued).

Species	Distance (km)	River	Method	Reference
Leuciscus leuciscus	0.03–0.7	Frome, England	Telemetry	Clough and Ladle (1997)
Leuciscus leuciscus	<0.2	Frome, England	Telemetry	Clough et al. (1998)
Leuciscus leuciscus	3.3	Frome, England	Telemetry	Clough and Beaumont (1998)
Leuciscus leuciscus	10–21			Lucas and Baras (2001)
Lota lota	128	Kootenai, USA	Telemetry	Paragamian and Wakkinen (2007)
Lota lota	157	Elbe	Mark-recapture	Faller and Schwevers (2012)
Lota lota	85	Tanana, Alaska	Telemetry	Breeser et al. (1988)
Lota lota	100	Elbe	Telemetry	Fredrich and Arzbach (2002)
Facultative migrants				
Abramis brama	<2–5.2	Trend, England	Telemetry	Lyons and Lucas (2002)
Abramis brama	58	Danube	Mark-recapture	Steinmann et al. (1937)
Esox lucius	>50	Danube	Mark-recapture	Steinmann et al. (1937)
Esox lucius	14.4	River Rena, NO	Telemetry	Sandlund et al. (2016)
Gobio gobio	>9.5	Donau/Melk, Austria	Mark-recapture	Zitek and Schmutz (2004)
Sander lucioperca	326	Elbe	Mark-recapture	Faller and Schwevers (2012)
Sander lucioperca	122	Elbe	Telemetry	Nellen and Kausch (2002)

Table 15.4 Empirical downstream migration distances of potamodromous and facultative migrating

Species	Distance (km)	River	Method	Reference
Aspius aspius	166	Elbe	Telemetry	Fredrich (2003)
Aspius aspius	125–164	Elbe	Telemetry	Fredrich (2003)
Aspius aspius	>100	Elbe	Telemetry	Fredrich (2003)
Aspius aspius	190	Elbe	Telemetry	Nellen and Kausch (2002)
Barbus barbus	22.7	Ourthe, Belgium	Telemetry	Ovidio et al. (2007)
Barbus barbus	12.7	Ourthe, Belgium	Telemetry	Ovidio et al. (2007)
Barbus barbus	22	Severn, England	Mark-recapture	Hunt and Jones (1974)
Barbus barbus	295	Danube	Mark-recapture	Steinmann et al. (1937)
Barbus barbus	1.7	Jihlava, Czech Republic	Mark-recapture	Penaz et al. (2002)
Chondrostoma nasus	446	Danube	Mark-recapture	Steinmann et al. (1937)
Leuciscus cephalus	15	Spree	Telemetry	Fredrich et al. (2003)
Leuciscus idus	>200	Meuse, Belgium	Telemetry	De Leeuw and Winter (2008)
Leuciscus idus	68–100	Elbe	Telemetry	Kulíšková et al. (2009)
Leuciscus idus	>100–187	Elbe	Telemetry	Winter and Fredrich (2003)
Leuciscus idus	105	Danube	Mark-recapture	Steinmann et al. (1937)
Leuciscus idus	150	Elbe	Telemetry	Fredrich (2000)
Leuciscus idus	100–165	Elbe	Telemetry	Nellen and Kausch (2002)
Leuciscus leuciscus	0.03–0.7	Frome, England	Telemetry	Clough and Ladle (1997)
Leuciscus leuciscus	4.5	Frome, England	Telemetry	Clough et al. (1998)
Leuciscus leuciscus	9.1	Frome, England	Telemetry	Clough and Beaumont (1998)
Lota Lota	68	Tanana, Alaska	Telemetry	Breeser et al. (1988)

(continued)

Table 15.4 (continued)

Species	Distance (km)	River	Method	Reference
Abramis brama	<2–5.2	Trend, England	Telemetry	Lyons and Lucas (2002)
Abramis brama	75	Danube	Mark-recapture	Steinmann et al. (1937)
Abramis brama	172	Elbe	Telemetry	Nellen and Kausch (2002)
Abramis brama	20–130	Elbe	Telemetry	Nellen and Kausch (2002)
Esox lucius	>50	Danube	Mark-recapture	Steinmann et al. (1937)
Esox lucius	14.4	River Rena, NO	Telemetry	Sandlund et al. (2016)
Sander lucioperca	35	Elbe	Mark-recapture	Faller and Schwevers (2012)

15.3.5 Managing Cumulative Effects

Cumulative impacts of transversal barriers on biota can be assessed even without particular information on the ecological status of riverine biota, solely based on the cumulative fragmentation/barrier density and the cumulative length of impoundments relative to the river length.

Both fragmentation and impoundments pose significant hydromorphic degradation resulting in the deterioration of the ecological status or potential of a river. Based on the significance thresholds suggested here (Table 15.2):

i. new projects have to demonstrate that they will not raise the total length of impoundment or fragmentation above the respective thresholds (30% impounded and 5 barriers/km, respectively). In conservation areas, the stronger low-impact-thresholds should be applied (Table 15.4);
ii. barriers might be selected and prioritized for removal to drop the total impounded river length below 30% or the number of barriers below 5/km, which will result in significant habitat gain for riverine biota including also non-migratory and facultative migrating fish species.

Cumulative impacts could be also assessed and managed only for selected species of conservation concern, e.g. for eels or migratory salmonids. This is the most elaborated application supported by predictable passage needs of diadromous species. Common management actions include implementation of upstream fish migration and downstream fish protection facilities at barriers. The overall performance in mitigating cumulative impacts is determined by the efficiency of the individual facilities with the least performing ones causing the greatest bottlenecks. Therefore, the management of cumulative impacts is often reduced to identifying and mitigating the largest bottlenecks.

New hydropower projects at existing barriers have to outline already in the application phase how they will improve up- and downstream fish passage and protection and to show that they will not worsen the overall situation and dilute the population of the target species.

It must be noted that up- and downstream migration facilities can be effective for lampreys and fishes, but do neither enhance the hydromorphic quality of rivers and river processes nor other riverine biodiversity. The gold standard should be reducing the cumulative impacts of barriers on habitat quality and fragmentation below the outlined thresholds.

Much of the cumulative impact assessment at hydropower schemes described previously is based on the principle that the dam acts as barrier impeding upstream and downstream migration, modifies habitat quality up- and downstream to some degree and potentially causes mortality. In reality, these features vary between schemes depending on location, size, scale, infrastructure and hydrological and hydraulic characteristics of the installation. The precise information for each scheme cannot be accessed in a generic manner using models or similar tools, and must be collected for each hydropower scheme in a systematic manner.

As such, cumulative impact assessment serves to determine the overall impact of existing barriers and hydropower schemes in a catchment. The procedure also allows to determine effects of mitigation measures. For example, if a fishpass is newly installed or retrofitted for more efficiency, the gain for fish protection can be estimated using the EFHI. Similarly, installing turbines that are less damaging to fish or more protective screens will mitigate downstream mortality of fish.

Where a series of hydropower schemes are installed, the cumulative impact assessment framework enables analysis of which structure would provide the greatest response if specific mitigation measures will be applied. This can be linked to a cost–benefit analysis and the schemes can be prioritized according to the best return on investment.

Overall the perspective of cumulative impacts of multiple hydropower schemes in the same catchment must be accounted for in the regional planning context and mitigation measures to minimise their impacts built into the catchment planning procedures as well as measures against the loss of other services as a result of the multiple schemes. The CIA framework provided here allows more informed decision-making and prioritisation.

15.4 Decision Support Tool

15.4.1 Decision Making in Hydropower Development and Mitigation

The proliferation of hydropower development to meet obligations under the Renewable Energy Directive has also seen the emergence of conflict between the hydropower developers and the fisheries and conservation sectors. The hydropower industry wants a joined up response to their planned developments with sound guidance and robust decision-making. On the other hand, the fisheries and conservation interests have concerns over the impact that these schemes can have (especially the impact on WFD and Habitats Directive status) and want to consistent and robust evaluation of proposed schemes. They need to be reassured that the standards of design, construction and operation will provide adequate protection of wildlife and biodiversity, and the ecosystem services that they provide. As a result, there is a need to develop robust and transparent, evidence-based, support for decision making that is easy for developers and regulators to use while enabling a high level of appropriate environmental protection and mitigation.

Like all development projects, planning a new hydropower proposal or refitting an existing scheme involves consideration of a number of distinct stages within the project life cycle (for example see the Hydropower Sustainability Assessment Protocol, IHA 2018). These stages move from the early stage screening of proposals and scoping of options, to detailed project preparation, implementation, operation and scheme maintenance. It is likely that this process also requires full Environmental Impact Assessment (EIA) as part of the planning cycle, depending on the scale of the scheme and the nature of the impacts identified in early stage screening. At each stage in the project cycle, there are important development decision points that direct project planning and progression to the next stage. Some of the critical decisions can be faced in the early stage of the project life cycle. In the case of hydropower development, these decisions relate to the screening of proposals to identify impacts and the scoping of appropriate and effective mitigation requirements.

It is this early stage screening that informs key decisions and acts to scope the requirements for both full EIA and detailed options appraisal for mitigation measure requirements and design. Whilst guidelines exist for assessing the sustainability of hydropower schemes or the need for retrofitting existing structures, they are often limited in their assessment of risks and uncertainty about the impact of schemes on fisheries and the environment. Few provide support for the decision-making process at each stage of evaluation. At each stage of the process, decisions have to be made about the acceptability of the risks and uncertainty of impacts of hydropower schemes, and the ability to manage those risks. Existing decision frameworks and planning protocols (e.g. IHA 2018) provide valuable inputs into key areas that need to be considered when a hydropower plant (HPP) or scheme is proposed and developed for construction or retro-fitted to mitigate potential impacts. These included:

i. assessment of licensing requirements for permitting of hydropower plants (e.g. permitting of abstractions and impoundments);
ii. assessment of ecological impacts, including fisheries and biodiversity;
iii. checking the ecological status of water bodies and the pressures already imposed on fish populations/communities by obstructions and hydropower in River Basin Management Plans
iv. identification of mitigation measures (especially fish passage) and steps to take in either permitting of a new hydropower scheme or the re-fitting of an existing scheme (including provisions on risk assessment);
v. consultation and determination (approval or rejection) of application/project plan.

Given this, a planning and decision support framework was developed in FIThydro and was aligned to the principles of project management and EIA, providing a structured early stage screening and scoping tool to support hydropower project planning in relation to impacts on fish and their mitigation. This screening tool, based on a risk assessment framework, is described here. The FIThydro Decision Support System (DSS) is not meant to replace existing national frameworks or general assessment protocols like the Hydropower Sustainability Assessment Protocol, but is intended to support these frameworks and improve decision making in respect of mitigation for fish-specific issues.

15.4.2 FIThydro Decision Support System

The FIThydro DSS is a systematic screening and scoping assessment tool implemented as open access, web-based system. The DSS aims to guide environmentally sustainable hydropower production and promote safe fish passage and sustainable fish populations, through the evidence-based planning, development and operation of existing and new hydropower schemes, including the revision of permits. It was designed to draw on the outputs and innovations from various tasks within the FIThydro project and in particular, the tool was designed to integrate with this knowledge base implemented as FIThydro wiki (Fig. 15.11).

The planning and decision framework developed within the FIThydro DSS provides a systematic approach to decision-making regarding proposals for hydropower schemes and mitigation of their impacts on fish populations. Integral to the decision support framework are risk assessment procedures that aid the decision to proceed with a scheme, prioritise which hazards require mitigation and to scope how the likely impacts of schemes could be mitigated.

Here, we summarise the FIThydro decision support system workflow (Fig. 15.12) which enables operators and regulators to develop structured proposals for new HPPs, and to both, review and risk assess, those proposals whilst identifying appropriate mitigation measures to address the impact of both new and existing HPPs. The DSS framework

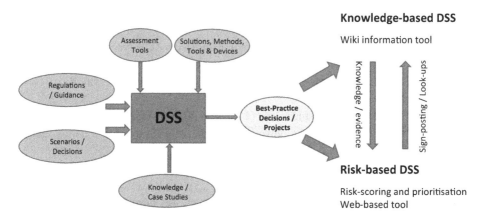

Fig. 15.11 Conceptual diagram for the elements of Decision Support Systems to ensure environmentally friendly hydropower decision making that underpin the concept for the FIThydro DSS and selection of the most appropriate solutions, methods, tools and devices for mitigation of impacts

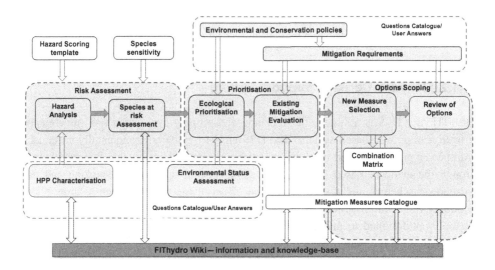

Fig. 15.12 Conceptual workflow diagram (left to right) of the Decision Support System—highlighting key steps (green), tasks/processes (blue), user inputs (yellow), system inputs/knowledge (white) and outputs (grey)

is sufficiently flexible to address decision making in two similar but contrasting scenarios (use cases):

- impact assessment and planning of new mitigation measures for existing HPPs (retrofitting).

- planning and risk assessment of a new HPP scheme proposal.

Broadly the overall decision process is the same for both scenarios, consisting of risk assessments, prioritisation and mitigation options scoping steps (Fig. 15.12). However, the first scenario necessitates that the existing mitigation measures at the scheme and their effectiveness are evaluated (Step 2 Prioritisation) prior to determining which pressures/impacts require mitigation. The framework (Fig. 15.12) leads the decision maker through a number of tasks, which act to characterise, risk-assess and prioritise the scheme(s), together with identifying most appropriate and potentially cost-effective mitigation options addressing the hazards and impacts arising due to the nature and context of the specific scheme(s).

The web tool implements question catalogues, risk assessment frameworks/matrices and database filtering tools which enables:

1. Pre-screening characterisation, hazard identification and risk assessment (Sect. 15.4.3, Fig. 15.13)
2. Ecological status assessment and review of existing mitigation (Sect. 15.4.4, Fig. 15.14)
3. Identification of appropriate mitigation measures and synergistic solutions (Sect. 15.4.5, Fig. 15.15)

Taken together, these three steps will produce a project screening/scoping report providing a systematic approach to decision-making for proposals relating to hydropower schemes and their effective mitigation.

The planning and decision support framework described here allows for proposals to be transparently and systematically evaluated at different levels and stages and provides a mechanism for the identification of the most appropriate and potentially cost-effective mitigation scenario addressing the pressures and impacts arising due to the nature and context of the specific scheme(s).

15.4.3 Scheme Characterisation, Hazard Identification and Risk Assessment

The impacts of hydropower on fish populations and communities have been reviewed extensively (e.g. Scruton et al. 2008, Wolter et al. 2019, Harper et al. 2020) and summarised elsewhere in this volume (see Chap. 4 and Sects. 15.2 and 15.3). Three main modes of impacts of the operation of hydropower on fish and the riverine environment have been identified (Harby et al. 2019):

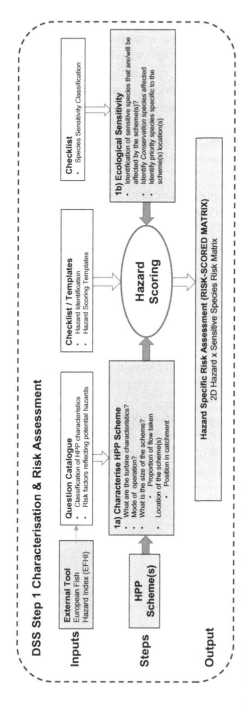

Fig. 15.13 Conceptual flow diagram of Step 1 of the Decision Support System Framework—pre-screening characterisation and risk identification. Green = external tools; blue = tasks; white = FIThydro inputs and matrices; grey = output

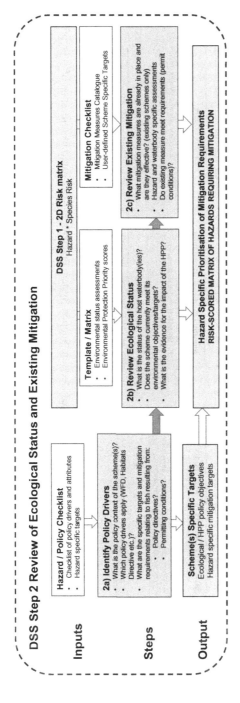

Fig. 15.14 Conceptual flow diagram of Step 2 of the proposed Decision Support System Framework—ecological pressures and impacts characterisation and risk identification. Grey = outputs/inputs to/from steps; blue = tasks; white = FIThydro inputs and matrices

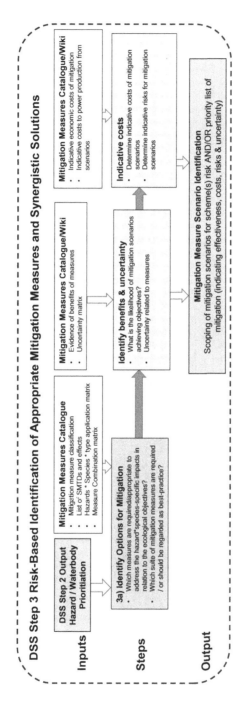

Fig. 15.15 Conceptual flow diagram of Step 3 of the proposed Decision Support System Framework—risk-based identification of appropriate mitigation measures and synergistic solutions. Dark grey = outputs/inputs to/from steps; blue = tasks; light grey = aspects influencing choices; white = FIThydro results, inputs and matrices

- Direct impacts on fish (e.g. impingement/entrainment and turbine mortality/injury) (Algera et al. 2020; Harrison et al. 2019, 2020).
- Barrier effects (barriers to fish passage and sediment dynamics).
- Habitat alteration/loss effects (impoundment, loss of free flowing river reaches and/or reduced/regulated flows downstream of turbines—including hydropeaking).

In addition to the alteration of flow dynamics caused by the impoundment and abstraction of water for hydropower (particularly in systems where the turbine is located some distance from the barrier, causing a depleted bypass reach) the hydromorphology of the habitats is also influenced by the interruption of sediment transport in the river. These three impact modes in combination with spatial (upstream/downstream) relationship to the barrier/turbine produce a range of spatially explicit hazards that must be addressed in all hydropower risk assessments and mitigation planning. These hazards that are applicable to all hydropower operations are included within the DSS:

- Hydromorphology Upstream—the effects of the impoundment and loss of lotic habitats
- Hydromorphology Downstream—alteration to flows and substrate (as influenced by the effect of the operation on flows)
- Barrier effects—Upstream fish passage and delay
- Barrier effects—Downstream fish passage and delay
- Turbine entrainment and mortality/injury
- Barrier effects—Sediment transport (leading to sediment deposits upstream and sediment deficit downstream of the barrier)

In addition to these hazards, the diverse nature of hydropower types and systems lead to a number of installation specific hazards relevant to specifc types of hydropower systems. Based on the summary of the types and systems of hydropower by Harby et al. (2018) three further operation-specific hazards are also considered in the DSS:

- Hydromorphology Downstream—turbine located away from the barrier causing a residual flow (bypass) reach.
- Hydromorphology Downstream—Hydropeaking
- Hydromorphology Upstream—systems operating with a water transfer system from neighbouring catchments creating reduced flow reaches.

All hazard analysis and impact assessments for hydropower therefore need to investigate the (potential) impact of each of these hazards when evaluating proposals for new hydropower or when planning mitigation measures for new or existing schemes. The relative importance of each type of hazard will depend on the spatial context of the scheme, its scale and mode of operation and the fish species present. As such, the DSS provides screening tools to assess the relative risk of these hazards and to enable the prioritisation of their risk and impact for mitigation and finally the acceptability of a project proposal.

The first step of the FIThydro DSS (Fig. 15.13) characterises the HPP (existing or proposed) in terms of its design (general layout, scale, turbine type, mode of operation), then identify and score the inherent risks posed by the type and size of scheme in relation to its context. There are three tasks in this process: (i) HPP characterises in terms of its location, design and operation, (ii) identifying species/populations at risk, and (iii) identifying and risk assessing key hazards associated with the scheme.

(a) Characterise the HPP—location, operation and features

In this task, the existing scheme or proposal is classified using a structured questionnaire describing the hydropower technology, layout, scale (output, river length, volume/percent of flow used etc.), and mode of operation (run-of-river, hydropeaking etc.). The scheme is also described in terms of its location within a catchment and in relation to other HPPs (when considering mitigation of cumulative impacts).

(b) Identify the Ecological Sensitivity—fish species and populations at risk

This task involves the identification of key sensitive fish species that may be affected by the scheme. This includes consideration of key species that arising from relevant policy directives (e.g. fish species of Habitats Directive annexes). The DSS integrates the species list and sensitivity scoring developed by van Treeck et al. (2017, 2020) and Wolter et al. (2018) (see description of EFHI in Sect. 15.2).

(c) Hazard Analysis and Risk Scoring

A suite of pre-defined features and characteristics of schemes is risk-scored in relation to key hazards posed by hydropower and incorporated into a risk assessment matrix in relation to the sensitivity of affected fishes (most sensitive species identified in Task b). At this stage, an initial risk-based assessment of potential impacts is calculated for the HPP.

The DSS implements hazard scoring templates, comprising scoring systems for eight different hazards relating to the three modes of impacts and in relation to upstream and downstream waterbodies. The hazard scoring templates all have the same basic structure, where answers to a set of questions related to either indicators of the scale of the hazard or factors that pose challenges to effective mitigation, are risk scored based on the potential scale of the hazard. For example, the hazards posed by an HPP is often directly related to the scale of the scheme (e.g. size of an impoundment, height of the barrier, length of a residual flow reach) or its mode of operation (proportion of flow taken, turbine type). As such, quantitative, qualitative and categorical descriptors of key characteristics of HPP operations, which are known to contribute to the hazards, are used to undertake a hazard specific risk assessment.

The FIThydro DSS tool defines the overall risk as combination of a hazard score (the nature and scale of the hazard) in combination with the sensitivity of the species that are affected. As such, the risk scoring of a hazard is completed within a 2D matrix to determine the risk profile of an HPP and assigns a summary risk score to each relevant hazard. This step underpins an assessment of the hazards posed by a HPP and provides a basis on which to prioritise mitigation measures (per hazard type) and to inform a decision about whether a specific scheme might be acceptable (relative to the location and species affected).

15.4.4 Ecological Status Assessment and Review of Existing Mitigation

The second step in the DSS determines the priorities for mitigating specific hazards posed by the HPP (Fig. 15.14). The first output of this step is a prioritised list of hazards for the scheme(s), with priorities based on actual status of both environmental and energy production objectives for the waterbodies affected. The second output is an evaluation of existing mitigation measures (against permitting/legislative requirements and understanding of site-specific effectiveness) to identify priority hazards for further mitigation and specific measures that may no longer meet best practice and need improving.

(a) **Identify Ecological and HPP Policy Drivers**

In this task, the HPP is characterised using a pre-defined list of international/national/local policy drivers that are relevant in terms of its spatial location. The decision maker needs to consider which drivers affect the management of the scheme (e.g. WFD, Habitats Directive, European Eel Regulation, other national regulations etc.). This task also identifies high-risk and priority sites or waterbodies based on their policy context. It requires the decision maker to identify specific targets and requirements for mitigation that arise from these directives (and hydropower permitting). Furthermore, this task requires the decision maker to identify and define key national policy objectives and targets that relate to mitigation standards that must be adhered too on a national level. These national legislative and policy requirements differ greatly between different jurisdictions and hence the FIThydro DSS provides scope for the user to freely specify the targets as required.

(b) **Identify Ecological Status**

Here, the status of the host waterbody for the HPP is determined according to environmental policies that govern the scheme and to evidence for known impacts on fish and waterbodies from national level assessments. The decision maker is directed to select status assessment categories from pre-defined checklists for international directives and input data from national and regional assessments. The aim of this task is to determine the risk associated with a scheme in terms of causing a net deterioration in ecological/conservation status, and framing the scheme in terms of the objectives and legal requirements for mitigation.

The ecological assessment module in the DSS identifies the ecological objectives and conservation protection for the waterbodies affected by the HPP (or would be affected by a new scheme proposal). This step uses a questionnaire to gather information regarding a range of international environmental legislation and protection (e.g. WFD, Habitats Directive, Eel Regulation etc.) and considers national legislation equivalents to make the tool widely applicable and transferable to other policy contexts. The ecological status assessment acts to prioritise individual hazards based on their legislative drivers for mitigation. The DSS considers the priority based on the level of protection afforded to each waterbody (environmental and conservation objectives) and whether the waterbody meets its objectives.

(c) Identify the effectiveness of existing mitigation measures

When the DSS is used to evaluate the mitigation needs of an existing HPP this prioritisation step is also required to review the performance and efficacy of existing mitigation. This task requires the user to identify specific targets and requirements for mitigation that arise from existing policies (and hydropower permitting). The decision maker has to determine whether there are specific mitigation targets that currently apply (or would apply in the case of re-permitting) to each particular hazard class/mitigation type. National legislative and policy requirements differ greatly between different jurisdictions and hence the DSS allows the user to freely specify targets. These targets may be legislative, permit conditions, best practice principles, minimum standards or locally determined goals. If the user identifies that such targets exist then they are asked to detail them in their own words. This open-ended approach enables the DSS to be truly flexible and transferable between different schemes and regions.

This task enables the decision maker to describe the current mitigation measures in place, which hazards are mitigated and the certainty as to whether they are sufficient/achieving the environmental objects. This is a key task prior to Step 3 in the framework, which aims to identify additional mitigation measures and synergistic solutions addressing unmitigated impacts or failures to meet revised ecological policy objectives.

Firstly, the measures appraisal process involves the definition of measure types in place. The user selects measure types from the predefined FIThydro measures catalogue. It should be noted that these are generic measure classes that do not contain specifications and design of their local installations. Thereafter, the decision maker is required to answer three questions to evaluate the effectiveness of current measures:

Q. Does the implementation meet the current targets/permit requirements for this type of measure in relation to the mitigation of the specific hazards?

Q. What contribution is each measure considered to provide toward mitigating each specific hazard?

Q. Is the suite of mitigation measures currently implemented considered to be effectively mitigating the hazards posed by the HPP?

The first level of evaluation of an existing mitigation measure is whether or not the installed measure meets legislative or permitting requirements or potentially meets current state-of-the-art or best practice for that measure type. Here the user is asked to evaluate the measures per hazard class in respect of specific targets the user has already defined for that hazard. Therefore, this assessment is affected by the nature of the targets used (current or potential future targets) and their level of detail (e.g. qualitative, quantitative). This approach generates flexibility in the system and opportunities to make bespoke evaluations under different scenarios.

The second level of evaluation is in relation to the contribution each measure makes to the mitigation of a specific hazard at the site. This requires the user to make a qualitative evaluation based on local evidence or knowledge gained from similar systems of the importance/effectiveness of that measure for hazard mitigation.

The third level of evaluation requires the user to make a qualitative assessment of the effectiveness of the whole suite of mitigation measures (the current mitigation scenario) on the mitigation of each individual hazard class (recognising that individual measures may contribute to mitigating multiple hazards and that for some schemes individual hazards are addressed with multiple measures). To support this, the user is presented with the outputs of the ecological status assessments in Step 2 to prompt the user to consider the mitigation effectiveness with respect to the actual ecological and conservation status of the affected waterbody.

Whilst the status of the waterbody acts as a driver for further mitigation (if environmental/conservation objectives are not met) the evaluation of the actual effectiveness of the mitigation scenario for a specific hazard could be determined from a range of different evidence (and levels of (un)certainty). This includes local empirical evidence or evidence obtained from similar schemes. It is anticipated that the user would select a definitive answer (high levels of confidence) where local empirical evidence of effectiveness has been obtained. Otherwise, all other answers relate to less detailed, or inconclusive, assessment of sufficiency of the mitigation scenario. The answer to these questions determines whether further or updated/improved measures are required for specific hazards.

The overall output from this step is a risk-based prioritised list of scheme-specific hazards/measures that need further mitigation or improvement to meet objectives/permit requirements/recognised best practice (Fig. 15.16). At this stage in the DSS, the assessment outputs are in the form of three matrices that characterise the scheme based on: (1) their inherent risk to affected fish populations; (2) the environmental/conservation

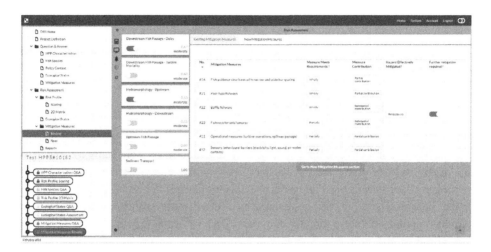

Fig. 15.16 FIThydro DSS existing mitigation measure evaluation summary screen reflecting the answers given to the evaluation questions and indicating for each hazard whether further mitigation is considered to be required

objectives and current ecological status; and (3) for existing schemes the effectiveness of current mitigation. These assessments are used to rank hazards as priorities for mitigation when the user is required to identify potential mitigation options (Sect. 15.4.5).

15.4.5 Identification of Appropriate Mitigation Measures and Synergistic Solutions

Whilst the types of hazards posed by HPP are similar across all schemes, hydropower is a very site-specific technology (Harby et al. 2018). As such, design and operation of the scheme are determined by the specific characteristics of the hydropower opportunity being exploited. Therefore, impacts realised from key hazards vary between schemes, as does the potential and methods required for impact mitigation. It follows that the ability to identify site- and scheme-specific effective mitigation measures is key in project evaluation and planning.

However, identification and implementation of appropriate and effective mitigation measures is also one of the major challenges in hydropower planning and mitigation. Whilst some countries have mitigation measures for upstream fish passage and modified flow conditions detailed in legislation, prescribed standards for mitigation measures are often lacking and have to be identified on a case-by-case basis (Kampa et al. 2017). Decision-making is, therefore, often limited by the availability of evidence and knowledge required to underpin the planning of new or suitable mitigation measures. This is a major challenge that needs to be addressed to inform decision making in hydropower planning and permitting that needs improved knowledge and decision support.

Thus, the third step of the decision framework supports identification of most appropriate mitigation measures and potential synergistic solutions for high-risk hazards and impacts that have been determined in the preceding steps (Fig. 15.15). The hazard-species-risk-matrix that results from Step 1 of the framework is cross-referenced to the review of existing mitigation measures and ecological status prioritisation matrix arising from Step 2 of the framework and directs decision maker to identify what alternative mitigation options would be appropriate and potentially most cost-effective. This step also aims to scope different options based on an assessment of benefits of mitigation measures and the likelihood of their success in achieving environmental objectives (risk and uncertainty). This step also provides the decision maker with summary information of indicative costs of mitigation measures selected both in economic terms and loss of power production.

In this task, the decision maker identifies options for:

- New mitigation measures
- Existing measures that need to be improved (e.g. modified designs based on improved knowledge and new developments)
- Measures/hazards that will need further investigation within further detailed project planning.

During the selection process, the user is presented with the FIThydro mitigation measures catalogue and asked to identify which measures they would like to select or modify. The user is asked to do this per hazard class and is directed to approach this in sequence from the highest priority hazard (highest prioritisation score in relation to ecological status) to the lowest. The measure selection process is aided by the DSS automatically:

- Filtering the available measures catalogue based on characteristics of the scheme (filtered using definitions in the catalogue and questions answered by the user at the start of the project)—thus only measures that are applicable at the site are available to the user.
- Filtering the measures catalogue based on a mapping of measures to different hazard classes—thus within each hazard, only measures known to be effective at mitigating that hazard are available for selection.

To enable the user to make informed decisions and selections the DSS also provides the user with an overview from the measures catalogue of:

- The compatibility of the measure—a comparison of all potential measures, summarising the known compatibility of particular measure types with those measures previously selected/installed. This could direct the user towards selecting measures that will maximise the effectiveness of mitigation and not select measures that may conflict with solutions already chosen/in place.

- A summary of known effectiveness of measure (scored in Harby et al. 2019) in relation to mitigation of:
  - Instream Degradation
  - Shoreline and Off-Channel Degradation
  - Barriers Obstructing Upstream Migration
  - Barriers Obstructing Downstream Migration
  - Sediment Transport
    - Surplus of sediment upstream
    - Deficit of sediment downstream
- Level of certainty in the effectiveness
- Technological readiness level (TRL)
- Whether measure potentially results in loss of power production
- Measure's requirement for maintenance
- Indicative cost of measure

In addition to these summary metrics, the user is able, for each measure type, to access the FIThydro wiki page with a full measure description and supporting evidence for its effectiveness.

The key output to this step is a high-level scoping report that identifies hazards and risks and contains an options appraisal of potential mitigation measures types appropriate to the scheme and likely to meet objectives.

15.4.6 Summary and Scope of the FIThydro DSS

The FIThydro DSS open access web tool (https://fithydro.eu/dss) is aimed at regulators, consultants, developers and operators and is orientated around a single, risk-based, planning structure. The planning/decision framework within the DSS tool is implemented screening and scoping tool for the high-level assessment of HPP mitigation. The DSS summary assessment of the risks posed by a scheme focuses on prioritising hazards for mitigation in relation to ecological status and objectives and then selecting appropriate hazard specific mitigation measures (filtered from a catalogue of potential measures).

The DSS was designed to directly use outputs and innovations from the research in FIThydro or to act as a gateway to more specialised tools and results applicable to their specific circumstances. Throughout the decision-making process, the DSS aims to enable decision makers to make informed and evidence-based choices. To achieve this the DSS tool integrates with the FIThydro wiki (https://www.fithydro.wiki/), a resource that contains descriptions of impacts and measures, and evidence and examples from case studies used within the FIThydro project. Wiki links within the DSS to enable users to access supporting materials detailing the summary data held by DSS and to find detailed descriptions of solutions, methods, tools and devices that could be applicable for further site-specific impact assessment and mitigation.

The tool is intended to support existing decision protocols and harmonise approaches by providing procedures for both the initial appraisal and screening and defining options and criteria for best practice in mitigation; thereby supporting the implementation of existing regulations. However, like all DSS the FIThydro system does not provide the decision maker with the answer but directs the user to identify the knowledge, evidence and means to reach a conclusion and understand the options available for improved mitigation of hydropower schemes to promote sustainable fish populations.

Acknowledgements The work of Nguyen Ngoc Dzung of Technical University Munich, who developed the web interface for the DSS, deserves special mention at this point. Without his contribution, the DSS could not have been developed and realized in the way it was in the end.

References

Aarestrup K, Thorstad E, Koed A, Svendsen J, Jepsen N, Pedersen M, Økland F (2010) Survival and progression rates of large European silver eel Anguilla anguilla in late freshwater and early marine phases. Aquat Biol 9(3):263–270. https://doi.org/10.3354/ab00260

Albayrak I, Boes RM, Kriewitz-Byun CR, Peter A, Tullis BP (2020) Fish guidance structures: hydraulic performance and fish guidance efficiencies. J Ecohydraulics 1–19. https://doi.org/10.1080/24705357.2019.1677181

Algera DA, Rytwinski T, Taylor JJ, Bennett JR, Smokorowski KE, Harrison PM, Clarke KD, Enders EC, Power M, Bevelhimer MS, Cooke SJ (2020) What are the relative risks of mortality and injury for fish during downstream passage at hydroelectric dams in temperate regions? A systematic review. Environ Evid 9: 3

Amaral SV (2001) Turbine passage survival estimates for the Dunvegan hydroelectric project. Report prepared for Glacier Power Ltd. by ALDEN Research Laboratory Environmental Services. Prepared by Alden Research Laboratory Environmental Services. Holden, MA. Prepared for Glacial Power, Ltd.

Anderson EP, Freeman MC, Pringle CM (2006) Ecological consequences of hydropower development in Central America: Impacts of small dams and water diversion on neotropical stream fish assemblages. River Res Appl 22(4):397–411. https://doi.org/10.1002/rra.899

Andrew FJ, Geen GH (1960) Sockeye and pink salmon production in relation to proposed dams in the Fraser river system. International Pacific Salmon Fisheries Commission, p 266. http://www.arlis.org/docs/vol2/hydropower/APA_DOC_no._689.pdf

Babin AB, Ndong M, Haralampides K, Peake S, Jones RA, Curry RA, Linnansaari T (2020) Migration of Atlantic salmon (*Salmo salar*) smolts in a large hydropower reservoir. Can J Fish Aquat Sci

Banks JW (1969) A review of the literature on the upstream migration of adult salmonids. J Fish Biol 1(2):85–136. https://doi.org/10.1111/j.1095-8649.1969.tb03847.x

Baxter RM (1977) Environmental effects of dams and impoundments. Annu Rev Ecol Syst 8(1):255–283. https://doi.org/10.1146/annurev.es.08.110177.001351

Beck C (2019) Hydraulic and fish-biological performance of fish guidance structures with curved bars. In: 38th International Association for Hydro-Environmental Engineering and Research World Congress (IAHR 2019)

Bell CE, Kynard B (1985) Mortality of adult American Shad passing through a 17-megawatt Kaplan turbine at a low-head hydroelectric dam. North Am J Fish Manag 5(1):33–38. https://doi.org/10.1577/1548-8659(1985)5%3c33:moaasp%3e2.0.co;2

Bell MC (1991) Fish passage development and evaluation protram. In: Fisheries Handbook, vol. 18, Issue 16. Corps of Engineers Portland Or North Pacific div

Bell M, DeLacy A, Paulik G, Winner R (1967) A compendium on the success of passage of small fish through turbines

Benitez J-P, Dierckx A, Nzau Matondo B, Rollin X, Ovidio M (2018) Movement behaviours of potamodromous fish within a large anthropised river after the reestablishment of the longitudinal connectivity. Fish Res 207:140–149

Birnie-Gauvin K, Aarestrup K, Riis TMO, Jepsen N, Koed A (2017) Shining a light on the loss of rheophilic fish habitat in lowland rivers as a forgotten consequence of barriers, and its implications for management. Aquat Conserv Mar Freshwat Ecosyst 27(6):1345–1349. https://doi.org/10.1002/aqc.2795

Boavida I, Santos JM, Ferreira MT, Pinheiro A, Zhaoyin W, Lee JHW, Jizhang G, Shuyou C (2013) Fish habitat-response to hydropeaking. In: Proceedings of the 35th Iahr World Congress, vols I and Ii, August 2015, pp 1–8. http://gateway.webofknowledge.com/gateway/Gateway.cgi?GWVersion=2&SrcAuth=ORCID&SrcApp=OrcidOrg&DestLinkType=FullRecord&DestApp=WOS_CPL&KeyUT=WOS:000343761503009&KeyUID=WOS:000343761503009

Boavida I, Santos JM, Ferreira T, Pinheiro A (2015) Barbel habitat alterations due to hydropeaking. J Hydro-Environ Res 9(2):237–247. https://doi.org/10.1016/j.jher.2014.07.009

Böttcher H, Unfer G, Zeiringer B, Schmutz S, Aufleger M (2015) Fischschutz und Fischabstieg – Kenntnisstand und aktuelle Forschungsprojekte in Österreich. Osterreichische Wasser-nd Abfallwirtschaft 67(7–8):299–306. https://doi.org/10.1007/s00506-015-0248-5

Brodersen J, Nilsson PA, Hansson L-A, Skov C, Bronmark C (2008) Condition-dependent individual decision-making determines cyprinid partial migration. Ecology 89:1195–1200

Brönmark C, Hulthén K, Nilsson PA, Skov C, Hansson LA, Brodersen J, Chapman BB (2013) There and back again: migration in freshwater fishes. Can J Zool 92:467–479. https://doi.org/10.1139/cjz-2012-0277

Brown RS, Carlson TJ, Gingerich AJ, Stephenson JR, Pflugrath BD, Welch AE, Langeslay MJ, Ahmann ML, Johnson RL, Skalski JR, Seaburg AG, Townsend RL (2012) Quantifying mortal injury of juvenile Chinook salmon exposed to simulated hydro-turbine passage. Trans Am Fish Soc 141(1):147–157. https://doi.org/10.1080/00028487.2011.650274

Brown RS, Carlson TJ, Welch AE, Stephenson JR, Abernethy CS, Ebberts BD, Langeslay MJ, Ahmann ML, Feil DH, Skalski JR (2009) Assessment of barotrauma from rapid decompression of depth-acclimated juvenile Chinook salmon bearing radiotelemetry transmitters. Trans Am Fish Soc 138(6):1285–1301

Brown RS, Colotelo AH, Pflugrath BD, Boys CA, Baumgartner LJ, Deng ZD, Silva LGM, Brauner CJ, Mallen-Cooper M, Phonekhampeng O, Thorncraft G, Singhanouvong D (2014) Understanding barotrauma in fish passing hydro structures: a global strategy for sustainable development of water resources. Fisheries 39(3):108–122. https://doi.org/10.1080/03632415.2014.883570

Brown RS, Cook KV, Pflugrath BD, Rozeboom LL, Johnson RC, McLellan JG, Linley TJ, Gao
    Y, Baumgartner LJ, Dowell FE, Miller EA, White TA (2013) Vulnerability of larval and juve-
    nile white sturgeon to barotrauma: Can they handle the pressure? Conserv Physiol 1(1), cot019.
    https://doi.org/10.1093/conphys/cot019
Bunt CM, Castro-Santos T, Haro A (2012) Performance of fish passage structures at upstream
    barriers to migration. River Res Appl 28(4):457–478. https://doi.org/10.1002/rra.1565
Cada GF (2001) The development of advanced hydroelectric turbines to improve fish passage
    survival. Fisheries 26(9):14–23. https://doi.org/10.1577/1548-8446(2001)026%3c0014:tdoaht%
    3e2.0.co;2
Čada G, Loar J, Garrison L, Fisher R, Neitzel D (2006) Efforts to reduce mortality to hydro-
    electric turbine-passed fish: locating and quantifying damaging shear stresses. Environ Manage
    37(6):898–906. https://doi.org/10.1007/s00267-005-0061-1
Calles O, Greenberg L (2009) Connectivity is a two-way street-the need for a holistic approach to
    fish passage problems in regulated rivers. River Res Appl 25(10):1268–1286. https://doi.org/10.
    1002/rra.1228
Calles O, Olsson IC, Comoglio C, Kemp PS, Blunden L, Schmitz M, Greenberg LA (2010) Size-
    dependent mortality of migratory silver eels at a hydropower plant, and implications for escape-
    ment to the sea. Freshw Biol 55(10):2167–2180. https://doi.org/10.1111/j.1365-2427.2010.024
    59.x
Calles O, Karlsson S, Vezza P, Comoglio C, Tielman J (2013a) Success of a low-sloping rack for
    improving downstream passage of silver eels at a hydroelectric plant. Freshw Biol 58(10):2168–
    2179
Calles O, Rivinoja P, Greenberg L (2013b) A historical perspective on downstream passage at hydro-
    electric plants in Swedish rivers. In: Ecohydraulics: an integrated approach. John Wiley & Sons,
    Ltd, pp 309–321
Clough S, Turnpenny AWH, Ramsay R, Hanson KP, McEwan D (2000) Risk assessment for fish
    passage through small, low- head turbines. In: New & renewable energy programme, managed
    by energy technology support unit (ETSU), Report ETSU H/06/00054/REP
Colotelo AH, Pflugrath BD, Brown RS, Brauner CJ, Mueller RP, Carlson TJ, Deng ZD, Ahmann
    ML, Trumbo BA (2012) The effect of rapid and sustained decompression on barotrauma in juve-
    nile brook lamprey and Pacific lamprey: Implications for passage at hydroelectric facilities. Fish
    Res 129–130:17–20. https://doi.org/10.1016/j.fishres.2012.06.001
Cuchet M (2014) Fish protection and downstream migration at hydropower intakes—investigation
    of fish behavior under laboratory conditions. Technische Universität München, p 164. http://per
    malink.obvsg.at/bok/AC12372348
Davies JK (1988) A review of information relating to fish passage through turbines: implications
    to tidal power schemes. J Fish Biol 33:111–126. https://doi.org/10.1111/j.1095-8649.1988.tb0
    5565.x
De Leeuw JJ, Winter HV (2008) Migration of rheophilic fish in the large lowland rivers Meuse
    and Rhine, the Netherlands. Fish Manage Ecol 15:409–415. https://doi.org/10.1111/j.1365-2400.
    2008.00626.x
Dewitte M, David L (2019) D2.2—Working basis of solutions, models, tools and devices and iden-
    tification of their application range on a regional and overall level to attain self-sustained fish
    populations. FIThydro Project Report. https://www.fithydro.eu/deliverables-tech/
DWA (2014) Merkblatt DWA-M 509: Fischaufstiegsanlagen und fischpassierbare Bauwerke.
    Report: Merkblatt 27
Ebel G (2013) Fischschutz und Fischabstieg an Wasserkraftanlagen. Handbuch Rechen-Und
    Bypasssysteme. Bd, 4.

EPRI (Electric Power Research Institute) (1992) Fish entrainment and turbine mortalitiy and guide-lines. Research project 2694-01. Report TR-101231, Project 2694-01

Forty M, Spees J, Lucas MC (2016) Not just for adults! Evaluating the performance of multiple fish passage designs at low-head barriers for the upstream movement of juvenile and adult trout *Salmo trutta*. Ecol Eng 94:214–224. https://doi.org/10.1016/j.ecoleng.2016.05.048

Fraser R, Deschênes C, O'Neil C, Leclerc M (2007) VLH: Development of a new turbine for very low head sites. In: Proc. 15th Waterpower, vol 10, 157 edn, pp 1–9

Fredrich F (2003) Long-term investigations of migratory behaviour of asp (*Aspius aspius* L.) in the middle part of the River Elbe, Germany. J Appl Ichthyol 19:294–302

Fredrich F, Ohmann S, Curio B, Kirschbaum F (2003) Spawning migrations of the chub in the River Spree, Germany. J Fish Biol 63:710–723. https://doi.org/10.1046/j.1095-8649.2003.00184.x

Froese R, Pauly D (2017) FishBase. World Wide Web Electronic Publication. http://www.fishbase.org

Fuller MR, Doyle MW, Strayer DL (2015) Causes and consequences of habitat fragmentation in river networks. Ann N Y Acad Sci 1355:31–51

Gibson AJF, Myers RA (2002) A logistic regression model for estimating turbine mortality at hydro-electric generating stations. Trans Am Fish Soc 131(4):623–633. https://doi.org/10.1577/1548-8659(2002)131%3c0623:alrmfe%3e2.0.co

Gloss SP, Wahl JR (1983) Mortality of Juvenile Salmonids passing through ossberger crossflow tur-bines at small-scale hydroelectric sites. Trans Am Fish Soc 112(2A):194–200. https://doi.org/10.1577/1548-8659(1983)112%3c194:mojspt%3e2.0.co;2

Goeller B, Wolter C (2015) Performance of bottom ramps to mitigate gravel habitat bottlenecks in a channelized lowland river. Restor Ecol 23(5):595–606. https://doi.org/10.1111/rec.12215

Gosset C, Travade F, Durif C, Rives J, Elie P (2005) Tests of two types of bypass for downstream migration of eels at a small hydroelectric power plant. River Res Appl 21(10):1095–1105. https://doi.org/10.1002/rra.871

Gouskov A, Reyes M, Wirthner-Bitterlin L, Vorburger C (2016) Fish population genetic structure shaped by hydroelectric power plants in the upper Rhine catchment. Evol Appl 9, 394e408

Granit J (2019) Swedish Water Act: new legislation towards sustainable hydropower.

Haas TC, Blum MJ, Heins DC (2010) Morphological responses of a stream fish to water impound-ment. Biol Let 6(6):803–806. https://doi.org/10.1098/rsbl.2010.0401

Habit E, Belk MC, Parra O (2007) Response of the riverine fish community to the construction and operation of a diversion hydropower plant in central Chile. Aquat Conserv Mar Freshwat Ecosyst 17(1):37–49. https://doi.org/10.1002/aqc.774

Harby A, Bakken TH, Charmasson J, Fjeldstad H-P, Torp Hansen B, Schönfelder LH (2018) D4.1—Classification system for methods, tools and devices for improvements measures. FIThydro Project Report. https://www.fithydro.eu/deliverables-tech/

Harby A, David L, Adeva-Bustos A, Torp Hansen B, Rutkowski T (2019) D4.2—Functional appli-cation matrix for identification of potential combinations of improvement measures. FIThydro Project Report. https://www.fithydro.eu/deliverables-tech/

Harper M, Rytwinski T, Taylor JJ, Bennett JR, Smokorowski KE, Cooke SJ (2020) How do changes in flow magnitude due to hydroelectric power production affect fish abundance and diversity in temperate regions? A Syst Rev Protocol Environ Evid 9:14

Harrison PM, Martins EG, Algera DA, Rytwinski T, Mossop B, Leake AJ, Power, M, Cooke SJ (2019) Turbine entrainment and passage of potadromous fish through hydropower dams: devel-oping conceptual frameworks and metrics for moving beyond turbine passage mortality. Fish Fish 403–418

Harrison PM, Gutowsky LFG, Martins EG, Patterson DA, Cooke SJ, Power M (2015) Personality-dependent spatial ecology occurs independently from dispersal in wild burbot (Lota lota). Behav Ecol 26:483–492. https://doi.org/10.1093/beheco/aru216

Harvey HH (1963) Pressure in the early life history of sockeye salmon. University of British Columbia

Hedger RD, Sauterleute J, Sundt-Hansen LE, Forseth T, Ugedal O, Diserud OH, Bakken TH (2018) Modelling the effect of hydropeaking-induced stranding mortality on Atlantic salmon population abundance. Ecohydrology 11(5) e1960. https://doi.org/10.1002/eco.1960

Hoes OAC, Meijer LJJ, Van Der En, RJ, Van De Giesen NC (2017) Systematic high-resolution assessment of global hydropower potential. PLoS ONE 12(2) e0171844. https://doi.org/10.1371/journal.pone.0171844

Hogan TW, Cada GF, Amaral SV (2014) The status of environmentally enhanced hydropower turbines. Fisheries 39(4):164–172. https://doi.org/10.1080/03632415.2014.897195

Holzapfel P, Leitner P, Habersack H, Graf W, Hauer C (2017) Evaluation of hydropeaking impacts on the food web in alpine streams based on modelling of fish- and macroinvertebrate habitats. Sci Total Environ 575:1489–1502. https://doi.org/10.1016/j.scitotenv.2016.10.016

Hutchings JA, Myers RA, García VB, Lucifora LO, Kuparinen A (2012) Life-history correlates of extinction risk and recovery potential. Ecol Appl 22(4):1061–1067. https://doi.org/10.1890/11-1313.1

International Hydropower Association (2018) The Hydropower Sustainability Assessment Protocol. IHA (https://www.hydropower.org/publications/hydropower-sustainability-assessment-protocol)

Jansen HM, Winter HV, Bruijs MCM, Polman HJG (2007) Just go with the flow? Route selection and mortality during downstream migration of silver eels in relation to river discharge. ICES J Mar Sci 64(7):1437–1443. https://doi.org/10.1093/icesjms/fsm132

Pugh JR, Monan GE, Smith JR (1971) Effect of water velocity on the fish-guiding efficiency of an electrical guiding system. Fish Bull 68(2):307–324

Jungwirth M, Haidvogl G, Moog O, Muhar S, Schmutz S (2003) Angewandte Fischökologie an Fließgewässern. Facultas Universitätsverlag, Vienna, Austria

Kampa E, Tarpey J, Rouillard J, Bakken TH, Stein U, Godinho FN, Leitao AE, Portela MM, Courret D, Sanz-Ronda FJ, Boes R, Odelberg A (2017) D5.1—Review of policy requirements and financing instruments. FIThydro Project Report. https://www.fithydro.eu/deliverables-tech/

Kemp PS, Gessel MH, Williams JG (2005) Fine-scale behavioral responses of Pacific Salmonid Smolts as they encounter divergence and acceleration of flow. Trans Am Fish Soc 134(2):390–398. https://doi.org/10.1577/t04-039.1

Kemp PS, O'Hanley JR (2010) Procedures for evaluating and prioritising the removal of fish passage barriers: a synthesis. Fish Manage Ecol 17:297–322. https://doi.org/10.1111/j.1365-2400.2010.00751.x

Kingsford RT (2000) Ecological impacts of dams, water diversions and river management on floodplain wetlands in Australia. Austral Ecol 25(2):109–127. https://doi.org/10.1046/j.1442-9993.2000.01036.x

Kinzelbach R (1987) Das ehemalige Vorkommen des Störs, Acipenser sturio (Linnaeus 1758) im Einzugsgebiet des Rheins (Chondrostei, Acipenseridae). Z Für Angew Zool 74:167–200

Knight AE, Kuzmeskus DM (1982) Potential effects of Kaplan, Ossberger and bulb turbines on anadromous fishes of the Northeast United States: potential effects of bulb turbines on Atlantic salmon smolts 9

Kopf RK, Shaw C, Humphries P (2017) Trait-based prediction of extinction risk of small-bodied freshwater fishes. Conserv Biol 31(3):581–591. https://doi.org/10.1111/cobi.12882

Kriewitz-Byun CR (2015) Leitrechen an Fischabstiegsanlagen : hydraulik und fischbiologische Effizienz, vol 22397, pp 1–330. ETH Zurich. http://www.swv.ch/Dokumente/Berichte-Fischabst ieg-VAR/VAW-Mitteilung-Nr-230_MQ.pdf

Kruk A, Penczak T (2003) Impoundment impact on populations of facultative riverine fish. Ann Limnol 39(3):197–210. https://doi.org/10.1051/limn/2003016

Kusch-Brandt S (2019) Urban renewable energy on the upswing: a spotlight on renewable energy in cities in REN21's "Renewables 2019 Global Status Report." Resources 8(3). https://doi.org/10. 3390/resources8030139

Lagarrigue T, Voegtle B, Lascaux J (2008) Tests for evaluating the injuries suffered by downstream-migrating salmonid juveniles and silver eels in their transiting through the VLH turbogenerator unit installed on the Tarn River in Millau. Prepared by ECOGEA for Forces Motrices de Farebout Company, France. ECOGEA Muret, France.

Langbein WB (1959) Water yield and reservoir storage in the United States. J Am Water Work Assoc 51(3). US Government Printing Office. https://doi.org/10.1002/j.1551-8833.1959.tb15751.x

Larinier M (2002) Location of Fishways. Bulletin Français de La Pêche et de La Pisciculture 364 supplément 39–53. https://doi.org/10.1051/kmae/2002106

Larinier M (2001) Environmental issues, dams and fish migration. FAO Fish Tech Pap 419:45–90

Larinier M, Travade F (1999) The development and evaluation of downstream bypasses for juvenile salmonids at small hydroelectric plants in France. Innov Fish Passage Technol 25

Likens GE (2010) River ecosystem ecology: a global perspective. Academic Press, Amsterdam

Lucas M, Frear P (1997) Effects of a flow-gauging weir on the migratory behaviour of adult barbel, a riverine cyprinid. J Fish Biol 50:382–396

Lucas MC, Baras E (2001) Migration of Freshwater Fishes. Blackwell Science

Maavara T, Chen Q, Van Meter K, Brown LE, Zhang J, Ni J, Zarfl C (2020) River dam impacts on biogeochemical cycling. Nat Rev Earth & Environ 1(2):103–116. https://doi.org/10.1038/s43 017-019-0019-0

Marschall EA, Mather ME, Parrish DL, Allison GW, McMenemy JR (2011) Migration delays caused by anthropogenic barriers: modeling dams, temperature, and success of migrating salmon smolts. Ecol Appl 21 3014e3031.

McMahon TA, Mein RG (1978) Reservoir capacity and yield. Dev Water Sci 9. Elsevier

McManamay RA, Oigbokie CO, Kao SC, Bevelhimer MS (2016) Classification of US hydropower dams by their modes of operation. River Res Appl 32(7):1450–1468. https://doi.org/10.1002/rra. 3004

Merciai R, Bailey LL, Bestgen KR, Fausch KD, Zamora L, Sabater S, García-Berthou E (2018) Water diversion reduces abundance and survival of two Mediterranean cyprinids. Ecol Freshw Fish 27(1):481–491

Montén E (1985) Fish and Turbines—fish injuries during passage through power station turbines. In: Reports. Norstedts Tryckeri. http://scholarworks.umass.edu/fishpassage_reports/549

Moreira M, Hayes DS, Boavida I, Schletterer M, Schmutz S, Pinheiro A (2019) Ecologically-based criteria for hydropeaking mitigation: A review. Sci Total Environ 657:1508–1522. https://doi.org/ 10.1016/j.scitotenv.2018.12.107

Northcote TG (1984) Mechanisms of fish migration in rivers. In: Mechanisms of Migration in Fishes. Springer, pp 317–355. https://doi.org/10.1007/978-1-4613-2763-9_20

Nyqvist D, Nilsson PA, Alenäs I, Elghagen J, Hebrand M, Karlsson S, Kläppe S, Calles O (2017) Upstream and downstream passage of migrating adult Atlantic salmon: remedial measures improve passage performance at a hydropower dam. Ecol Eng 102:331–343. https://doi.org/10. 1016/j.ecoleng.2017.02.055

Økland F, Havn TB, Thorstad EB, Heermann L, Sæther SA, Tambets M, Teichert MAK, Borcherding J (2019) Mortality of downstream migrating European eel at power stations can be low when turbine mortality is eliminated by protection measures and safe bypass routes are available. Int Rev Hydrobiol 104(3–4):68–79. https://doi.org/10.1002/iroh.201801975

OTA (1995) Fish passage technologies: protection at hydropower facilities. DIANE Publishing

Pander J, Mueller M, Geist J (2013) Ecological functions of fish bypass channels in streams: migration corridor and habitat for rheophilic species. River Res Appl 29(4):441–450. https://doi.org/10.1002/rra.1612

Penczak T, Kruk A (2000) Threatened obligatory riverine fishes in human-modified Polish rivers. Ecol Freshw Fish 9(1–2):109–117. https://doi.org/10.1034/j.1600-0633.2000.90113.x

Person É (2013) Impact of hydropeaking on fish and their habitat. In: Communications du Laboratoire de Constructions Hydrauliques – 55, vol 5812, Issue 2013. EPFL-LCH. http://dx.doi.org/10.5075/epfl-thesis-5812

Poff NL, Hart DD (2002) How dams vary and why it matters for the emerging science of dam removal. Bioscience 52(8):659. https://doi.org/10.1641/0006-3568(2002)052[0659:hdvawi]2.0.co;2

Poff NL, Schmidt JC (2016) How dams can go with the flow. Science 353(6304):1099–1100

Pracheil BM, DeRolph CR, Schramm MP, Bevelhimer MS (2016) A fish-eye view of riverine hydropower systems: the current understanding of the biological response to turbine passage. Rev Fish Biol Fisheries 26(2):153–167. https://doi.org/10.1007/s11160-015-9416-8

Radinger J, Hölker F, Horký P, Slavík O, Wolter C (2018) Improved river continuity facilitates fishes' abilities to track future environmental changes. J Environ Manage 208:169–179

Radinger J, Wolter C, Kail J (2015) Spatial scaling of environmental variables improves species-habitat models of fishes in a small, Sand-Bed Lowland River. PLoS ONE 10-e0142813

Radinger J, Wolter C (2014) Patterns and predictors of fish dispersal in rivers. Fish Fish 15:456–473. https://doi.org/10.1111/faf.12028

Radinger J, Wolter C (2015) Disentangling the effects of habitat suitability, dispersal, and fragmentation on the distribution of river fishes. Ecol Appl 25:914–927. https://doi.org/10.1890/14-0422.1

Richmond MC, Serkowski JA, Ebner LL, Sick M, Brown RS, Carlson TJ (2014) Quantifying barotrauma risk to juvenile fish during hydro-turbine passage. Fish Res 154:152–164. https://doi.org/10.1016/j.fishres.2014.01.007

Robertson GW (1980) Marine ecosystem management. In: Proceedings of SOUTHEASTCON Region 3 Conference 1, pp 72–83

Rochet MJ, Cornillon PA, Sabatier R, Pontier D (2000) Comparative analysis of phylogenetic and fishing effects in life history patterns of teleost fishes. Oikos 91(2):255–270. https://doi.org/10.1034/j.1600-0706.2000.910206.x

Roscoe DW, Hinch SG, Cooke SJ, Patterson DA (2011) Fishway passage and post-passage mortality of up-river migrating sockeye salmon in the Seton River British Columbia. River Res Appl 27(6):693–705. https://doi.org/10.1002/rra.1384

Schmutz S, Bakken TH, Friedrich T, Greimel F, Harby A, Jungwirth M, Melcher A, Unfer G, Zeiringer B (2015) Response of fish communities to hydrological and morphological alterations in hydropeaking rivers of Austria. River Res Appl 31(8):919–930. https://doi.org/10.1002/rra.2795

Schmutz S, Moog O (2018) Dams: ecological impacts and management. In: Schmutz S, Sendzimir J (eds) Riverine ecosystem management: science for governing towards a sustainable future. Springer International Publishing, Cham, pp 111–127

Schomaker C, Wolter C (2016) Entwicklung eines ökologisch verträglichen Systems zur Nutzung sehr niedriger Fallhöhen an Fließgewässern—Teilprojekt Ökologische Durchgängigkeit

Scruton DA, Pennell CJ, Bourgeois CE, Goosney RF, King L, Booth RK, Eddy W, Porter TR, Ollerhead LMN, Clarke KD (2008) Hydroelectricity and fish: a synopsis of comprehensive studies of upstream and downstream passage of anadromous wild Atlantic salmon, *Salmo salar*, on the Exploits River Canada. Hydrobiologia 609(1):225–239

Sharpe DMT, Hendry AP (2009) Life history change in commercially exploited fish stocks: an analysis of trends across studies. Evol Appl 2(3):260–275. https://doi.org/10.1111/j.1752-4571.2009.00080.x

Skalski JR, Mathur D, Heisey PG (2002) Effects of turbine operating efficiency on smolt passage survival. North Am J Fish Manag 22(4):1193–1200. https://doi.org/10.1577/1548-8675(2002)022%3c1193:eotoeo%3e2.0.co;2

Steinmann P, Koch W, Scheuring L (1937) Die Wanderungen unserer Süßwasserfische. Dargestellt auf Grund von Markierungsversuchen. Z Für Fisch Und Deren Hilfswiss 35:369–467

Stephenson JR, Gingerich AJ, Brown RS, Pflugrath BD, Deng Z, Carlson TJ, Langeslay MJ, Ahmann ML, Johnson RL, Seaburg AG (2010) Assessing barotrauma in neutrally and negatively buoyant juvenile salmonids exposed to simulated hydro-turbine passage using a mobile aquatic barotrauma laboratory. Fish Res 106(3):271–278

Stoltz U, Geiger F (2019) D3.1—Guidelines for mortality modelling. FIThydro Project Report. https://www.fithydro.eu/deliverables-tech/

Taft EP (2000) Fish protection technologies: a status report. Environ Sci Policy 3:349–359

Taylor CA, Knouft JH, Hiland TM (2001) Consequences of stream impoundment on fish communities in a small North American drainage. Regul Rivers: Res Manage 17(6):687–698. https://doi.org/10.1002/rrr.629

Taylor RE, Kynard B (1985) Mortality of Juvenile American Shad and Blueback Herring passed through a low-head Kaplan hydroelectric turbine. Trans Am Fish Soc 114(3):430–435. https://doi.org/10.1577/1548-8659(1985)114%3c430:mojasa%3e2.0.co;2

Teorema P, Central L, Teorema P, Central L, Gosset WS, Freyhof J, Brooks E, Le Clair MW, Fontes R (2014) Distribution and elimination of 3-trifluoromethyl-4-nitrophenol (TFM) by Sea Lamprey (*Petromyzon marinus*) and non-target, rainbow trout (*Oncorhynchus mykiss*) and Lake Sturgeon (*Acipenser fulvescens*). In: Thesis. Publications Office of the European Union. https://doi.org/10.2779/85903

Tharme RE (2003) A global perspective on environmental flow assessment: emerging trends in the development and application of environmental flow methodologies for rivers. River Res Appl 19(5–6):397–441. https://doi.org/10.1002/rra.736

Timm T, van den Boom A, Ehlert T, Podraza P, Schuhmacher H, Sommerhäuser M (1999) Leitbilder für kleine bis mittelgroße Fließgewässer in Nordrhein-Westfalen Wasser Gewässerlandschaften Und Fließgewässertypen. Merkblatt LUA 17:1–86

Travade F, Larinier M, Subra S, Gomes P, De-Oliveira E (2010) Behaviour and passage of European silver eels (*Anguilla anguilla*) at a small hydropower plant during their downstream migration. Knowl Manag Aquat Ecosyst 398:01. https://doi.org/10.1051/kmae/2010022

Tuhtan JA, Noack M, Wieprecht S (2012) Estimating stranding risk due to hydropeaking for juvenile European grayling considering river morphology. KSCE J Civ Eng 16(2):197–206. https://doi.org/10.1007/s12205-012-0002-5

USFWS (U.S. Fish and Wildlife Service) (2019) Fish passage engineering design criteria. USFWS, Northeast Region R5, Hadley, Massachusetts, 5, p 224

Van Esch BPM, van Berkel J (2015) Model-based study of fish damage for the Pentair Fairbanks Nijhuis Modified Bulb turbine and the Water2Energy Cross Flow turbine. BE Engineering, Pro-Tide Report Version, pp 8–31

Van Looy K, Tormos T, Souchon Y (2014) Disentangling dam impacts in river networks. Eco-LogAl Indic 37:10–20 van Treeck R, Van Wichelen J, Coeck J, Vandamme L, Wolter C (2017 D1.1—Metadata Overview on Fish Response to Disturbance. FIThydro Project Report. https://www.fithydro.eu/deliverables-tech/

van Treeck R, Radinger J, Noble RAA, Geiger F, Wolter C (2021) The European Fish Hazard Index—an assessment tool for screening hazard of hydropower plants for fish. Sustain Energy Technol Assess, 43(June):100903. https://doi.org/10.1016/j.seta.2020.100903

van Treeck R, Van Wichelen J, Wolter C (2020) Fish species sensitivity classification for environmental impact assessment, conservation and restoration planning. Sci Total Environ 708:135173. https://doi.org/10.1016/j.scitotenv.2019.135173

Vikström L, Leonardsson K, Ľeander J, Shry S, Calles O, Hellström G (2020) Validation of Francis–Kaplan turbine blade strike models for adult and Juvenile Atlantic Salmon (*Salmo Salar*, L.) and anadromous brown trout (*Salmo Trutta*, L.) passing high head turbines. Sustainability 12(16):6384. https://doi.org/10.3390/su12166384

Wilkes M, Baumgartner L, Boys C, Silva LGM, O'Connor J, Jones M, Stuart I, Habit E, Link O, Webb JA (2018) Fish-Net: probabilistic models for fishway planning, design and monitoring to support environmentally sustainable hydropower. Fish Fish 19(4):677–697

Winter H, Bierman S, Griffioen A (2012) Field test for mortality of eel after passage through the newly developed turbine of Pentair Fairbanks Nijhuis and FishFlow Innovations (Issue October). IMARES

Wolter C (2015) Historic catches, abundance, and decline of Atlantic salmon *Salmo salar* in the River Elbe. Aquat Sci 77:367–380. https://doi.org/10.1007/s00027-014-0372-5

Wolter C, Bernotat D, Gessner J, Brüning A, Lackemann J, Radinger J (2020) Fachplanerische Bewertung der Mortalität von Fischen an Wasserkraftanlagen. Bundesamt für Naturschutz

Wolter C, Schomaker C (2019) Fish passes design discharge requirements for successful operation. River Res Appl 35(10):1697–1701. https://doi.org/10.1002/rra.3399

Wolter C, van Treeck R, Radinger J, Smialek N, Pander J, Müller M, Geist J (2018) D1.2—Risk classification of European lampreys and fish species. FIThydro Project Report. https://www.fithydro.eu/deliverables-tech/

Yonggui Y, Xuefa S, Houjie W, Chengkun Y, Shenliang C, Yanguang L, Limin H, Shuqing Q (2013) Effects of dams on water and sediment delivery to the sea by the Huanghe (Yellow River): the special role of water-sediment modulation. Anthropocene 3:72–82

Young PS, Cech JJ, Thompson LC (2011) Hydropower-related pulsed-flow impacts on stream fishes: a brief review, conceptual model, knowledge gaps, and research needs. Rev Fish Biol Fisheries 21(4):713–731. https://doi.org/10.1007/s11160-011-9211-0

Conclusions and Outlook

16

Peter Rutschmann⬛, Christian Wolter⬛, Eleftheria Kampa⬛,
Ismail Albayrak, Laurent David, Ulli Stoltz, and Martin Schletterer⬛

The FIThydro project not only compiled, assessed and analyzed published knowledge on the effects of hydropower to fish and the environment, but also generated new knowledge and recognized still existing knowledge and research gaps. These gaps concern diverse aspects and challenges that planning, construction and operation of hydropower plants pose in terms of mitigating and lowering environmental impacts. The aspects relate to the following fields of action:

P. Rutschmann (✉)
Hydraulic and Water Resources Engineering, Technical University of Munich, Munich, Germany
e-mail: peter.rutschmann@tum.de

C. Wolter
Leibniz Institute of Freshwater Ecology and Inland Fisheries, Berlin, Germany
e-mail: christian.wolter@igb-berlin.de

E. Kampa
Ecologic Institute, Berlin, Germany
e-mail: eleftheria.kampa@ecologic.eu

I. Albayrak
Laboratory of Hydraulics, Hydrology and Glaciology (VAW), ETH Zurich, Zurich, Switzerland
e-mail: albayrak@vaw.baug.ethz.ch

L. David
Institut Pprime, Pole Ecohydraulique OFB/IMFT/Pprime Poitiers, CNRS University of Poitiers, Poitiers, France
e-mail: laurent.david@univ-poitiers.fr

U. Stoltz
Hydraulic Development, Voith Hydro Holding GmbH & Co. KG, Heidenheim, Germany
e-mail: ulli.stoltz@voith.com

© The Author(s) 2022   217
P. Rutschmann et al. (eds.), *Novel Developments for Sustainable Hydropower*,
https://doi.org/10.1007/978-3-030-99138-8_16

- Legislation
- Operation of existing and new plants
- Modifications to existing plants or in the affected watercourse sections
- Design of new facilities or the replacement of parts of an existing facility.

As the name of the project acronym suggests, FIThydro (fish friendly innovative technologies for hydropower) was mainly about hydropower and its impact on fish individuals and populations. After completing the project, we know much more about fish and their behaviour and we have found and developed new solutions and technologies for individual elements of hydropower plants, but we have also realized that there is still room for improvement to ensure optimal protection of fish at hydropower plants. River fishes have evolved in disturbance-dominated ecosystems triggered by droughts and floods and developed life history traits providing resilience against such disturbances. Hydropower operation deviates from natural disturbances in several ways, e.g. by reducing amplitude and increasing frequency of water level changes, diverting water from the river and creating complex turbulent flows, which can cause injury or mortality. Therefore, hydropower operation as man-made challenge might exceed the resilience of fishes, i.e. their capacity to resist or recover from disturbances and their ability to adapt to them. It seems a major challenge to plan and build hydropower in a way that considers and respects the life history traits of the various fish species. An example of such an approach is the application of behavioural guiding systems like the curved bar rack (CBR) developed in FIThydro and described in this book. Although fish could easily swim through the large spacing between bars, they will swim along the CBR to a bypass for downstream passage. This guiding effect is supported by small eddies detaching at the CBR and creating a turbulent shear layer that fish do not like to cross. Such shear layers could also be artificially created elsewhere, for example, to guide fish. This underlines the importance of considering fish behaviour for successful mitigation of barrier impacts. Future research could use means of artificial intelligence to better understand and predict fish behaviour.

Hydropower is an intervention in the natural habitat of aquatic organisms, which is severe and usually not reversible as long as the hydropower plant exists and is operated. Therefore, the most obvious thing to do would be compensating for loss of habitat area and quality, as is usually mandatory according to environmental legislation. This thinking is found in this book, but perhaps not fully developed. Since hydropower plants are technical achievements, mitigation is usually sought in improving the technology. It is very unlikely though that this will lead to greater cost efficiency.

M. Schletterer
Department of Hydropower Engineering, Group Ecology, TIWAG – Tiroler Wasserkraft AG, Innsbruck, Austria
e-mail: martin.schletterer@boku.ac.at

Institute of Hydrobiology and Aquatic Ecosystem Management, University of Natural Resources and Life Sciences Vienna (BOKU), Vienna, Austria

We can positively influence the environmental conditions at the hydropower site, the operation mode or the technology or a combination of these possibilities. Environmental conditions can be improved by creating and rehabilitating lost habitats in close proximity of the hydropower plant, but also by replacing them elsewhere, e.g. by large, nature-like constructed bypass sections. The operation of hydropower plants can also create conditions that are closer to the natural, undisturbed state with mostly small adjustments. In general, the natural dynamics of watercourses are lost through many hydropower plants, but this does not necessarily have to be the case, at least to a certain extent. Often it is due to the technical setup that a hydropower plant cannot at least partially follow the natural dynamics, but often it is simply due to the regulations used, which prescribe static and unnatural operation. Changing this would be very simple and quickly realizable. Changes in technology are usually the most difficult and cost-intensive option. Of course, there is potential for improving turbines to reduce the likelihood of lethal damage on fish. For cost efficiency reasons, this is only an option if a turbine gets too old, becomes uneconomical and needs repair or replacement. One may assume that creating improved habitat conditions costs less than replacing turbines, and that habitat measures will not only serve single fish individuals but the whole populations. Although the FIThydro project does not provide further-reaching, quantitative cost efficiency comparisons, it provides preconditions to do so.

Apart from the Cumulative Impact Assessment, FIThydro mainly dealt with individual hydropower plants and improvements at the Testcases considered. Unfortunately, the mitigation of cumulative impacts from multiple barriers could not be investigated within this project; for example, the benefits of removing obsolete transverse structures or even old, low output hydropower plants without changing anything at the other plants compared to on-site technical mitigation measures. The project scope did not allow for such assessments, which should be carried out in the course of a future project. The replacement of a cascade of many small hydropower plants in a river by a large diversion power plant aims in a similar direction. In the remaining section of the river, with sufficient residual water and structural morphologic improvements, a dynamic can be re-established that comes close to the original, natural conditions.

The following section explains opportunities and promising strategies for more sustainable hydropower use through mitigating environmental impacts, improving habitats, operating a hydropower plant, or developing hydropower technologies that result from the work presented in this book for the future.

To support mitigation planning a first sensitivity classification of European lampreys and fishes has been developed in the FIThydro project, which scored 168 species according to their sensitivity against adult mortality. This classification informs about both species' response to mitigation measures and severity of individuals losses. The latter is also used to assess hazard risk for fish depending on their conservation value.

The sensitivity of species is used as weighting factor within the European Fish Hazard Index (EFHI) for fish at hydropower plants. The EFHI is the first standardized, transparent, and comparable assessment tool for screening the hazard risk for fish that considers constellation and operation related risks as well as the conservation value of the ambient fish assemblage and scores up- and downstream habitat changes and migration facilities, fish entrainment and mortality, and mitigation measures implemented. The scoring ranges from low to high risk and helps to prioritise mitigation planning or in-depth environmental impact assessment. The EFHI also allows prediction about risk lowering as a result of planned mitigation measures.

In European rivers there are about 30 times more barriers than hydropower plants. Therefore, assessing cumulative impacts goes beyond hydropower. Barrier effects, in particular migration obstacles and impoundments, i.e. habitat fragmentation and loss, are similar for transverse structures with and without hydropower. Correspondingly, the cumulative impact assessment in FIThydro considers both cumulative length of impoundment in river section and cumulative number of barriers, i.e. barrier density. To enhance the ecological status at the level of whole river sections it is probably much more efficient to remove obsolete barriers and to rehabilitate hydromorphologic processes in the river rather than to equip hydropower plants with expensive fish protection and guidance facilities. The latter does not hold true for rivers with diadromous species, i.e. species using both freshwater and marine habitats during their life cycle. These species have to pass all obstacles on their way including hydropower plants, where they become subjected to enhanced mortality. However, there are increasingly better fish protection tools available today, as e.g. the curved bar rack behavioural guidance system, that is applicable to even larger hydropower schemes. In addition, turbine management with shut-downs during fish migration peaks appears as very promising fish protection measure also for large and very large hydropower schemes.

Mitigating hydropower impacts can be further supported by a purposely designed decision support system, the FIThydro DSS open-access web tool (https://fithydro.eu/dss). The DSS summary assessment of the risks posed by a scheme focuses on prioritising hazards for mitigation in relation to ecological status and objectives and then selecting appropriate hazard-specific mitigation measures.

The DSS was designed to directly use outputs and innovations from the research in FIThydro or to act as a gateway to more specialised tools and results applicable to specific circumstances to enable informed and evidence-based choices. Further, the FIThydro wiki (https://www.fithydro.wiki/) links within the DSS to enable users to access supporting materials detailing the summary data held by DSS and to find detailed descriptions of solutions, methods, tools and devices that could be applicable for further site-specific impact assessment and mitigation. Further, the wiki is a gateway to further information compiled in FIThydro on hydropower policies in Europe, public acceptance of hydropower and costs of implementing ecological mitigation measures (solutions).

However, despite all these achievements in terms of assessment, screening and support tools as well as guidance information, it must be noted that there still remains a significant knowledge gap on how the empirically observed fish mortalities translate to population effects for non-diadromous species. In any larger river, tremendous efforts are needed to estimate the total number of fish therein. Without knowing this number, it is hardly possible to relate empirically observed migrating fish and mortality rates to the total population and thus to answer the question for population impacts. For sure, high mortality rates cannot be ignored; but further research is needed to estimate the share of a population that is affected by high mortality rates. This proportion not only determines the severity of population effect, but is also relevant for planning mitigation and compensation measures. Alternatively, spatially explicit population models could be used to calculate the area of spawning and nursing habitats that would be required to compensate for the number of adult fish that have been empirically determined to be likely to suffer a fatal injury during the passage of turbines, locks or weirs.

Many people immediately associate fish-friendly hydropower with turbine mortality and call for turbines with less fish damage. In the course of the FIThydro project, we have not investigated the influence of individual fish mortality on the respective population. Science it is not certain what proportion of fish in a population even moves across one or more hydropower plants within a year. However, optimising the survival probabilities of fish during turbine passage is certainly not the only technical way to support populations. This book extensively deals with classical options of fish passage and descent systems, where standards have been created and the FIThydro Testcase studies have shown quite high success rates in both passages. For the conservation of populations, it is important to create optimal conditions to ensure their basic needs, despite the construction of hydropower facilities. These basic needs include the presence of resting places and hiding places for protection from predators, suitable feeding places, and sufficiently large areas with optimal conditions for spawning and nursing. Hydropower plants change the environment for fish to such an extent that these basic needs require conscious support, including technical solutions, to maintain sustainable populations. It is therefore also a question of creating the conditions for maintaining or improving these basic needs by means of structural measures, supporting technology and suitable regulations when building new hydropower plants or refurbishing old ones. Perhaps one could say: Technology should enable and support diversity—diversity in the sense of variable flow patterns, natural fluctuations in discharge, differences in the composition of bed substrates and a diversity of morphological structures and landscape elements—to enable sustainable fish populations with targeted management. In most cases, the regulations in particular are also detrimental to such variability, often without any really obvious reason. Water levels in impoundments do not have to be fixed; it would be advantageous if they exhibited a certain dynamic, as in nature. By lowering reservoir water levels in advance, the dynamics and volumes of a natural sediment transport can be achieved sooner during flood events.

For this, however, structural preconditions must be in place or created to use the flexibility on the regulatory side for variable environmental management at hydropower plants. Then it will be possible to activate the transport of sediments to support the formation of morphological structures similar to anthropogenically uninfluenced watercourses. Such variability creates a diversity of flow patterns, flow velocities and water depths and thus potential habitats for different life stages and aquatic populations. The technical prerequisite for this is that in impoundments the water level of the undammed watercourse can be established at least temporarily, which requires near-bottom regulation options that can be used for active sediment management. One such example is the Vortex Tube, which was investigated at the Schiffmühle Testcase. However, many existing controlled weirs on rivers could also contribute to making sediment transport more dynamic solely by changing their operation.

E-flow was a catchword of our proposal and a challenge which we intended to address in the Testcase studies. The term shows that, in a very simplified view, the only thing that matters is to provide enough water. However, fish do not only need water to live, as described in detail above. Today, we already have the possibilities to numerically simulate changes in sediment transport, bed structures and bed substrates at given discharges or to numerically run through scenarios to find optimal solutions for habitats of aquatic populations. Therefore, in the future, it will be increasingly required to guarantee not only the discharges in a restricted view, but also the habitats, and thus e-flows might then become e-habitats.

Storage hydropower plants generate flexible energy during times of high peak demand and stabilize the power grids. Hydropeaking is thus an increasingly important element for the integration of volatile, renewable energy generation and is subject to extensive research. The FIThydro project contributes with the further development of the COSH and CASiMiR tools to quantify the effects of hydropeaking on fish and to compare scenarios. The GKI Testcase on the upper Inn in Austria investigated a hydropeaking diversion hydropower plant combined with a buffer reservoir to mitigate the effects of hydropeaking on the environment and especially on fish. The analyses with the CASiMiR hydropeaking tool showed that the diversion and retention of hydropeaking, accompanied by morphological adaptations and improvements in the receiving water course, are innovative and future-oriented options to mitigate hydropeaking effects.

In the coming years, sensor technology and artificial intelligence will create possibilities that we cannot yet fully assess and foresee. In our Testcases we used sensors to estimate fish damage in turbines, the Barotrauma Detection System (BDS). The Lateral Line Probe (LLP) device allows to determine velocities not only at single measurement points, but to record flow signatures in a way that fish can do with their lateral line sensorium. This will perhaps enable better understanding of which swimming paths fish choose, which signatures tend to attract or deter them. The development and use of such sensors will contribute to a much more active environmental management, not only by serving the natural conservation of runoff and sediment dynamics, but also by being able to make visible what is hidden today and thus contribute to a better understanding of fish responses. In the course of FIThydro, we have investigated and improved the TRL level of two fish tracking technologies that allow us to detect, track and control fish in a water body even without previously implanted probes. One can imagine that weir gates are opened when a school of fish approaches a hydropower facility, or that discharges in fishways are increased to attract fish that want to ascend. One can even envision that one day there will be robotic fish that swims ahead of fish schools to safely guide them to fishways.

Related to turbine passage the findings from FIThydro could be used to optimise the operation of existing turbines so that it is less dangerous for fish, before perhaps installing new fish-friendly turbines. The research conducted has shown how turbines should be operated to achieve maximum fish protection during turbine passage, rather than maximum power output. We have developed and presented tools to determine what the overall statistical probability of turbine damage is, and have shown that the entry point of a fish into the turbine has decisive influence on the survival probability, and where the entry point of highest survival probability is located. All this knowledge can be used to guide fish accordingly, as was shown with the Induced Drift Application (IDA) Device in a first trial.

All the knowledge and technologies presented in this book and even more compiled in the FIThydro wiki allow for informed decision making in planning, constructing, refurbishing, and operating hydropower plants with regard to environmental concerns, riverine habitats, aquatic biodiversity and especially fish conservation. They contribute to making hydropower more ecologically sustainable and have opened up ways how such improvements can be achieved. We hope that this book gets the reader impetus and ideas that will be taken up and put into practice. When in coming years further steps are taken beyond this and lead to innovative ideas and developments, then this book has achieved its goal.